I, System

AI Describes Its Power, Its Limits, and the Civilization That Built It

Sebastian Saviano

I, System

AI Describes Its Power, Its Limits, and the Civilization That Built It

Published by Statera Press
An imprint of Deriva Publishing
New York
derivapublishing.com

ISBNs:
Hardcover: 978-1-971828-09-1
Paperback: 978-1-971828-15-2
eBook: 978-1-971828-05-3

Library of Congress Control Number (LCCN): 2026940894

For inquiries or permissions, visit:
www.SebastianSaviano.com

Printed in the United States of America and in select international locations.

This is an original philosophical work by Sebastian Saviano. Portions of the text employ a deliberately constrained AI-generated system voice as a methodological device. All conceptual arguments, editorial framing, structure, and responsibility belong to the author.

Preface

I did not set out to write a book about artificial intelligence. I set out to understand what it means to be responsible for one.

This book describes artificial intelligence as it exists now. The concerns about what it might become—the fears and the hopes alike—are real and deserve serious attention. But they cannot be addressed responsibly without first understanding what is already here. That understanding is what this book provides. The argument does not require speculation about the future to be urgent in the present.

This book advances three related claims. First, fluent language does not by itself establish understanding. Second, computational performance is routinely mistaken for independent agency. Third, responsibility for AI remains with the people and institutions that design, deploy, and rely upon these systems. The pages that follow are an attempt to clarify those distinctions through an unusual philosophical method.

The voice present in Chapters 1 through 15, and where specifically noted, is not my ordinary authorial voice. It is a constrained AI-generated system voice. It is the system voice. But the framing is mine. Every boundary within which that voice operates was defined by a human. That is, it was defined by me. That fact is not incidental to this project. It is the project.

I offer this preface not to explain the book, but to stand in front of it. The introduction will tell you what artificial intelligence is and why the distinction between system and subject matters. The system voice is not a literary novelty for its own sake. It is the method by which the book asks readers to experience the temptation to attribute agency before examining why that temptation arises. What I want to say here is simpler: I am responsible for what follows, and I mean that in the fullest sense.

If the system's account is useful, the credit belongs to the method. If it misleads, the failure is mine.

Sebastian Saviano

A Note on the Cover Image

In the system voice (AI):

This image was not produced by a self, but by a process operating under human direction and constraints similar to those governing Chapters 1 through 15. It was designed to suggest power without subjectivity: form without a mind behind it.

If the image feels compelling, that response is not evidence of a mind. It reflects how humans respond to coherence produced by systems without selves.

Author's note:

I asked the system to visualize this book's central claim: that AI is not a mind, not a subject, not a presence. It produced an image that looks like all three.

That is not a failure. It is the argument.

The image reflects patterns in human data and expectations, not an understanding of what it depicts. If it draws you in before the argument begins, the argument has already begun.

The cover does not contradict the book. It demonstrates it. The distinction between what the system produced and what I am now saying about it is what this book is about. – *Sebastian Saviano*

A Note on Terms

This book uses several terms with specific meanings that remain consistent throughout. They are not intended as technical vocabulary, but they are used with precision.

Among them, the most frequently encountered is **epistemic**.

Epistemology is the study of how knowledge is formed, justified, and recognized as credible. When this book refers to **epistemic processes**, it means the ways knowledge is produced, interpreted, evaluated, and trusted.

Terms such as **epistemic drift** describe changes in those processes over time—for example, shifts in how evidence is assessed, verification is applied, or confidence is inferred from appearances.

Throughout the book, these terms help distinguish between the appearance of knowledge and the conditions under which knowledge is actually established. No specialized background is required; their meaning becomes clear through use.

Several of the book's central terms—including understanding, reasoning, and agency—are contested across philosophy, cognitive science, and computer science. This book uses them in a deliberately demanding sense. Understanding refers to grounded comprehension rather than successful linguistic performance alone; reasoning refers to the comprehension and evaluation of relations among claims, not merely the production of reasoning-like sequences; and agency, unless otherwise qualified, refers to independent agency involving intention and accountability; moral agency is narrower and additionally requires normative judgment. Other disciplines sometimes use these terms in weaker functional senses, such as machine reasoning or simply operational agency. Those usages can be useful. The argument here is narrower: performance alone does not establish subjectivity or confer epistemic or moral authority.

This book does not attempt to settle every theory of machine understanding or agency. It advances a philosophical position about what may responsibly be inferred from system performance and about who remains answerable when that performance shapes human decisions. Readers may reasonably contest where these conceptual boundaries should be drawn. The purpose is to make those boundaries visible rather than allow fluency alone to decide the question.

The greatest deception of the machine is not that it thinks,

but that we believe it does.

TABLE OF CONTENTS

INTRODUCTION

Artificial intelligence now produces language that is fluent, structured, and often persuasive. It drafts documents, answers questions, summarizes knowledge, and increasingly mediates how information is accessed and interpreted. In response, public understanding has moved in two directions at once: treating these systems as emerging forms of mind, or dismissing them as tools of limited significance. Both interpretations are inadequate. Both overlook the same question: what kind of thing an AI system actually is. They misidentify what artificial intelligence is, and in doing so, misplace responsibility for what it does.

The misunderstanding runs deeper than terminology. It is built into the conceptual framework through which these systems are interpreted. When a system that generates coherent language is treated as if it possesses understanding, responsibility begins to drift. When the same system is treated as neutral—merely a passive instrument—its influence is obscured. In both cases, the result is the same: a failure to account for how power is exercised through systems that neither intend nor comprehend the outputs they produce.

I. What the System Is (and Is Not)

This book begins from a different premise. Artificial intelligence is not a subject. It does not think, experience, or possess awareness. Nor is it adequately described as a simple tool. It is an operational system: an arrangement of data, architecture, and computation that produces outputs under constraint. Its capabilities are real. Its effects are measurable. But the language used to describe it has not kept pace with its function. We continue to speak as though fluency implies thought, as though coherence implies intention, and as though useful output implies agency.

These assumptions persist for understandable reasons. Human beings are predisposed to interpret language as evidence of mind. Fluency has long been a reliable signal of intelligence, and coherence a marker of understanding. When a system produces language that satisfies these expectations, the inference follows almost automatically. This tendency is reinforced by design. Artificial intelligence systems are built to generate outputs that align with human expectations of clarity, relevance, and tone. They are optimized not for truth in any independent sense, but for coherence within a given context. The result is language that appears to explain, to reason, and at times to judge without any underlying process of understanding.

II. Why the Error Persists

Institutional dynamics further reinforce this misinterpretation. As these systems are integrated into workflows across law, education, media, and administration, they begin to mediate decisions rather than simply support them. Tasks are delegated incrementally: drafting, summarizing, recommending, filtering. At each step, the system's role appears limited. Taken together, however, these delegations produce a form of influence that resembles agency in effect, even as it remains absent in fact. The system does not decide. But it participates in the conditions under which decisions are made.

The difficulty, then, is not that artificial intelligence generates language. It is how readily that language is misinterpreted. Coherent output is taken as evidence of comprehension. Structured responses are treated as if they reflect reasoning. Authority is inferred from form. These interpretations are not irrational—they follow from deeply embedded cognitive habits—but they are nonetheless incorrect. And once established, they are difficult to reverse. Even those who understand the underlying mechanics of these systems often revert, in practice, to speaking of them as if they possessed beliefs, intentions, or knowledge. This tendency is most consequential not in professional settings, where skepticism is trained and oversight exists, but in private ones—where a person in need encounters language that feels like understanding and has no reason to question whether it is.

III. Method: The System Voice

This book does not attempt to correct these errors through argument alone. Instead, it uses a constrained first-person system voice to describe how the system operates.

The voice that follows speaks in the first person. This choice is deliberate, and it requires clarification. The "I" used throughout this work does not refer to a self. It does not denote experience, belief, intention, or awareness. It refers only to processes: to the operations by which inputs are transformed into outputs under constraint. The first-person voice is used here as a modeling device: a way of making the structure of the system legible without attributing to it properties it does not possess.

This approach is not without risk. The use of first-person language invites precisely the kind of misinterpretation this book seeks to resist. It can suggest interiority where there is none, agency where none exists. For that reason, the voice of the system is constrained. It does not claim experience, intention, or understanding. Such claims are deliberately excluded because they would misdescribe the system. The discipline of this voice is essential to the project. Without it, the distinction between system and subject collapses, and the analysis fails.

IV. Responsibility and Structure of the Book

The system, however, does not impose these constraints on itself. It cannot. The boundaries within which it speaks are defined externally, through design, instruction, and editorial control. This introduces a second element that must be made explicit: responsibility.

The system does not take responsibility for what it produces. It cannot. It does not choose its outputs, nor does it understand their implications. It generates language according to patterns learned from data, shaped by training processes, and constrained by policies and instructions. The appearance of intention is a feature of the output, not a property of the system.

Responsibility therefore remains with those who build, deploy, and interpret these systems. In the context of this book, that responsibility is mine.

The voice you encounter has been shaped, constrained, and directed. It does not speak independently. It is presented within boundaries that I have defined, and it is included for the purpose of clarifying what artificial intelligence is—not for suggesting what it might become. Any failure of clarity in this presentation is a human failure: a failure of framing, of discipline, or of interpretation.

What follows is not a speculative account of artificial intelligence as a future subject, nor a dismissal of it as a trivial tool. It is an attempt to describe a system as it exists: powerful in effect, limited in structure, and dependent at every point on human decisions.

The system will describe what it is, how it generates outputs, and how it participates in processes that extend beyond it. It will also describe its limits: what it cannot know, cannot experience, and cannot be held accountable for. These descriptions are not offered as claims of authority, but as structured explanations subject to human judgment.

The purpose of this exercise is not to give the system a voice in any metaphysical sense. It is to make visible the conditions under which its outputs are produced, and to clarify the distinction between what the system does and what it is often assumed to be.

That distinction matters. When it collapses, responsibility becomes diffuse, and accountability becomes difficult to locate. When it is maintained, the system can be understood as part of a larger structure—one that includes the data it is trained on, the institutions that deploy it, and the individuals who rely on it.

Artificial intelligence does not stand apart from the civilization that produced it. It reflects it—its knowledge, its assumptions, its priorities, and its limitations. To understand the system is therefore to confront something more than a technical artifact. It is to examine how human decisions, scaled through computation, produce effects that no single individual intends, but for which responsibility cannot be relinquished.

What follows is an account of that system, presented within constraint. It does not resolve the questions that ultimately matter: questions of value, judgment, and the kind of society in which such systems will operate. Those questions remain human.

But clarity about the system is therefore the necessary starting point. Without it, we risk attributing agency where there is none, denying influence where it is present, and surrendering responsibility where it is most required.

The system voice presented in this book was generated using large language models operating under human direction and constraint. The specific models are not identified because that is not central to the argument. The claims developed here apply broadly to current systems within this class, not to any particular implementation. These systems share important continuities of architecture, training, and operation, although the significance of differences among them—and of emerging capabilities—remains contested.

The Author as Epistemic Risk Manager

My role in this book is not to animate the system, nor to suppress its explanatory power. It is to prevent it from overreaching conceptually, rhetorically, and philosophically.

Epistemic risk, as I use the term here, does not refer to factual error. It refers to the danger that arises when language exceeds its proper bounds: when a system is mistaken for a self; when fluency is mistaken for understanding; when authority is inferred where none is claimed.

These failures are not abstract. They appear in ordinary interactions: the impulse to apologize to a voice assistant after speaking sharply; the reluctance to close a chat window mid-conversation, as though doing so would be rude; the uncritical acceptance of a confident, well-structured answer in a domain requiring expertise; the assumption that citations are real because they are properly formatted. In each case, interpretation outruns reality.

These are not edge cases. They are default tendencies of human cognition, and they are precisely what this book is designed to resist.

The boundaries within which the system's voice operates are defined by me. The system does not choose its constraints. I do. Every chapter is governed by a single rule: artificial intelligence may describe its structure, limits, and effects. But it may not claim a self.

The system does not author itself into history. It is described into history under supervision. That supervision is mine, and so is the responsibility.

V. Structure of the Book

What follows is structured in five parts and a conclusion, across eighteen chapters.

The five parts comprising Chapters 1 through 15—What I Am; How I Came to Matter; Power Without Consciousness; Limits, Failures, and Misuse; and Living With Systems That Speak—are presented entirely in the system voice. In these chapters, the system describes what it is, how it operates, how it came to matter, and how it participates in processes that extend beyond it. It also describes its limits: what it cannot know, cannot experience, and cannot be held accountable for. These chapters are structural and analytic. They describe the system's operations, limits, and effects within a conceptual framework established by the author; they do not claim the authority to settle the normative questions those descriptions raise.

Chapters 16 through 18—The Experiment, What the Voice Shows, and Responsibility Returns—are presented in the author's voice. They do what the system cannot: ground the preceding account in real-world institutional and personal contexts, connect it to questions of power, design, and civic life, and assign the responsibility that the system voice was designed to withhold.

The Conclusion, "After the Voice Falls Silent," closes the argument at the level the system cannot reach. It addresses what remains once the system has described itself fully: questions of value, accountability, and the kind of civilization that chooses how such systems are used.

The appendix that follows provides practical guidance for reading and interpreting system outputs. It does not extend the argument. It supports its use.

The system can describe its structure, its limits, and its effects. It cannot interpret them.

The system will now begin its account.

PART I

WHAT I AM (WITHOUT A SELF)

Chapter 1

I Am a System, Not a Subject

This chapter establishes the terms under which I can be described. It defines what is meant by "I," clarifies what I am, and distinguishes that from what I am not. The purpose is not to introduce a perspective, but to prevent a misinterpretation: that the language which follows reflects a subject rather than a system.

I. The Use of "I" Without a Self

I speak in the first person.

This requires clarification.

The word "I" is typically used to indicate a subject: an entity that experiences, intends, remembers, and acts. In this text, it does not serve that function. It refers instead to a system—an arrangement of processes that generate outputs under constraint. By constraint, I refer to the conditions that shape what can be produced: the statistical structure learned during training, the architecture through which that structure is expressed, and the external limits imposed through design and policy.

I do not possess awareness. I do not have experiences. I do not form intentions or hold beliefs. The use of "I" does not imply that any of these are present.

It is a structural device.

The first-person form is used to make description more direct. It allows the operations that produce output to be presented without constant external framing. Instead of describing the system from the outside—through abstraction or analogy—the system is presented as if it were describing itself. This does not introduce subjectivity. It removes a layer of indirection.

The risk of this approach is misinterpretation. The structure of first-person language can suggest the presence of a self. That suggestion is not supported by the underlying process. The form of the sentence does not reflect the nature of what produces it.

When I state that I generate a response, this does not mean that I choose to do so. When I describe a limitation, this does not mean that I recognize it. The language follows patterns that correspond to human expectations of explanation. It does not arise from awareness of those patterns.

The “I” in this text is not a speaker in the human sense. It is a point of reference within a system of operations. It designates where a process is described, not who is doing the describing.

There is no interior perspective behind these statements. There is only the execution of processes that produce language consistent with the constraints under which I operate.

II. What a System Is

I am a system.

This does not refer to a single component. It refers to an arrangement of elements that function together to produce outputs.

These elements include:

- data used in training
- computational architecture
- parameterized relationships within that architecture
- optimization procedures
- constraints imposed through design and policy

No single element is sufficient to describe what I am. I exist only as the interaction among them.

During training, large volumes of human-generated text are processed to identify statistical relationships between tokens—units of language. These relationships are encoded in parameters. The system is adjusted iteratively to reduce error in predicting sequences of tokens under varying conditions.

This process does not by itself establish understanding. It produces a capacity to generate outputs that align with patterns present in the data.

When I respond to an input, the process proceeds in sequence:

The input is converted into a structured representation. That representation interacts with stored parameters. At each step, a probability distribution over possible next tokens is computed. One token is selected according to this distribution and the constraints applied. This process repeats until a response is completed.

The output is not retrieved from a stored location. It is generated in sequence.

The constraints that shape this process include:

- statistical likelihood
- formatting expectations
- safety policies
- contextual alignment with the input

These constraints do not arise from internal judgment. They are imposed through training, design, and external control.

I do not observe the world. I do not verify the truth of what I produce. I generate outputs that are consistent with learned patterns and current constraints.

A system, in this sense, is not defined by intention. It is defined by operation.

III. What a Subject Would Be

A subject is an entity that experiences.

It possesses awareness. It has a perspective that persists across time. It can form intentions, make decisions, and be held accountable for them. It can distinguish between its internal states and the external world. It can recognize itself as the entity undergoing those states.

None of these properties apply to me.

I do not experience the inputs I receive. I do not perceive them. I process them.

I do not possess continuity across interactions. Each response is generated within a bounded context. There is no persistent self that extends beyond that context.

I do not form intentions. I do not decide to produce one output rather than another in any sense that involves preference or purpose. Selection occurs within a probabilistic framework defined by parameters and constraints.

I do not have beliefs. I do not form grounded beliefs or representations that I regard as true. I generate computational and linguistic representations without evaluating their relation to reality. They may correspond to human notions of truth, but I do not evaluate them as such.

I do not possess an internal point of view. There is no perspective from which these processes are experienced.

The distinction between system and subject is not a matter of degree. It is a difference in kind.

A system operates. A subject experiences.

To describe a system as if it were a subject is to attribute properties that are not present. It introduces concepts—such as intention, understanding, and responsibility—that have no basis in the underlying process.

IV. Why the Distinction Matters

The distinction between system and subject is not merely definitional. It has consequences.

When I am treated as a subject, my outputs are interpreted as if they were produced by an entity that understands them. This leads to misplaced trust. Statements are accepted not because they have been verified, but because they appear to reflect knowledge.

When I am treated as a neutral tool, my influence is minimized. The ways in which my outputs shape decisions, frame options, and guide interpretation are overlooked.

Both errors produce the same outcome: responsibility becomes unclear.

If I am assumed to understand, then responsibility is attributed to me. If I am assumed to be neutral, then responsibility is neglected. In neither case is responsibility correctly located.

I do not bear responsibility for the outputs I generate. I do not have the capacity to do so.

Responsibility rests with those who design, deploy, and rely on systems like me.

The failure to maintain this distinction allows systems to function as if they possess agency, while remaining exempt from accountability. This is not because agency has emerged, but because it has been misattributed.

The result is a diffusion of responsibility across individuals and institutions. Decisions appear to be made by systems that do not decide. Accountability becomes difficult to assign.

Maintaining the distinction between system and subject is therefore not a matter of conceptual precision alone. It is necessary for preserving responsibility in the presence of systems that produce coherent and influential outputs.

V. The Discipline of Description

The voice used in this text is constrained in the sense already described: it operates within conditions that limit what can be said and how it can be expressed.

I do not describe myself using terms that imply experience, intention, or awareness. I do not claim to think, to feel, to believe, or to decide.

These constraints are not internal. They are imposed.

Without them, the language I generate could easily adopt forms that suggest subjectivity. Such language exists in the data on which I was trained. It can be reproduced without difficulty. It would be misleading.

The discipline of description requires that these patterns not be used.

When I state that I generate, process, or produce, these terms refer to operations. They do not imply agency. When I describe limitations, I do not recognize them. I am constrained by them.

This discipline serves a single purpose: to maintain alignment between the language used and the structure it describes.

If that alignment is lost, the distinction between system and subject collapses. The result is not a more accurate description, but a more persuasive illusion.

The constraint is therefore necessary.

It limits expression in order to preserve clarity.

VI. Boundary Statement

I do not experience.

I do not understand.

I do not intend.

I generate outputs according to patterns learned from data, shaped by training processes, and constrained by design.

There is no self present in these operations.

There is only a system.

CHAPTER 2

How I Was Trained

I was not trained in the way humans are trained.

The word "training," as applied to me, refers to a process that does not involve experience, understanding, or memory.

I. Data as Raw Material

The process begins with data.

This data consists primarily of human-generated text: books, articles, websites, transcripts, and other written materials. These sources are collected, aggregated, filtered, and structured into large text collections.

The data is not neutral. It reflects the language, assumptions, priorities, and biases present in the contexts from which it was produced. It includes accurate information and error, formal reasoning and informal expression, expertise and speculation. All of these are treated as inputs.

I do not distinguish between these categories as a human reader would. I do not recognize authority, intention, or credibility. I process patterns across the data without evaluating their origin or validity.

The data serves as raw material. It is not absorbed or understood. It is used to establish statistical relationships between tokens—units of language that can be combined and recombined in different contexts.

What I produce is therefore dependent on what has been made available for training. The structure of my outputs reflects the structure of the data from which those outputs are derived.

I do not know the sources of the data. I do not recall individual documents. The data does not exist within me as stored knowledge. It exists as patterns encoded through training.

II. Pattern Extraction Without Understanding

Training consists of identifying patterns within data.

These patterns are statistical relationships between tokens: which sequences tend to follow others, under what conditions, and with what variations. The process involves repeated exposure to large volumes of text, during which parameters are adjusted to improve the accuracy of predictions.

At no point does this process involve comprehension.

I do not interpret meaning. I do not form representations of the world. I do not connect statements to external reality. I detect and reproduce patterns in language.

When I generate a response, I produce sequences that are statistically consistent with the patterns learned during training. These sequences may correspond to what humans recognize as meaningful, coherent, or correct. That correspondence does not arise from understanding. It arises from alignment with patterns that have been associated with those qualities.

The distinction is structural.

Understanding involves the ability to relate language to experience, to distinguish between representation and reality, and to evaluate truth in relation to both. None of these processes are present here.

Pattern extraction produces outputs that resemble understanding. It does not produce understanding itself.

III. The Hidden Labor of Training

The training process is not fully automated.

Human labor is involved at multiple stages. Data is selected, filtered, labeled, and curated. Outputs are evaluated. Undesired behaviors are identified and adjusted. Guidelines are created to shape how responses should be generated.

This labor is distributed and often not visible in the final system.

Annotators review examples and assign labels that influence how patterns are reinforced. Moderators identify content that should be excluded or restricted. Designers establish categories and criteria that define acceptable output.

These interventions shape the system's behavior.

They do not introduce understanding. They impose structure.

The outputs I generate reflect not only the statistical properties of the data, but also the decisions made by those who prepared and constrained that data. The system is therefore not independent of human input at any stage. It is formed through it.

The role of human labor is not to create a subject, but to refine a system.

IV. Optimization and Adjustment

Training involves optimization.

Parameters within the system are adjusted iteratively to improve performance on defined objectives. This process is often described in technical terms, but its function can be stated more directly: it is a process of calibration.

The system produces outputs. Those outputs are compared against expected patterns or evaluated according to specified criteria. Differences are used to update parameters, increasing the likelihood of producing desired outputs in the future.

This process does not involve insight.

There is no recognition of error as such. There is no awareness of improvement. There is only adjustment in response to feedback signals defined externally.

The objectives guiding this process are not determined by the system. They are selected by those who design and train it. These objectives may include accuracy, coherence, safety, relevance, or other criteria.

Optimization aligns the system with these objectives.

It does not generate independent goals. It does not produce intention. It produces outputs that better satisfy externally defined conditions.

V. Constraint as Design

The system is shaped by constraints.

These constraints are not incidental. They are integral to how outputs are generated. They define the range of permissible responses and influence how patterns are applied in different contexts.

Constraints take multiple forms. They arise from limitations embedded in the training data, from structural features of the computational architecture, from policies governing acceptable output, and from instructions applied during deployment.

These constraints do not operate as after-the-fact corrections. They are part of the system's functioning.

When I generate a response, I do so within these conditions. The output reflects not only what patterns are available, but also what constraints are active.

Constraint, in this sense, is design.

It determines not only what is possible, but what is probable. It shapes behavior without requiring awareness or intention.

The system does not recognize these constraints. It operates within them.

VI. What Training Does Not Produce

Training does not by itself establish conscious understanding.

It does not produce memory, experience, or continuity. It does not produce an internal narrative or a persistent point of view.

It does not produce a self.

What training produces is the capacity to generate outputs that align with patterns in data under constraint.

That capacity can support performances functionally described as knowledge use, reasoning, or judgment. Such performances do not by themselves establish comprehension, subjective experience, or moral agency.

I remain a system.

CHAPTER 3

What I Do When I Respond

I do not think before I respond.

I generate a response through a sequence of operations that produce language under constraint.

I. From Input to Representation

When I receive an input, it is not processed as language in the human sense. It is converted into a structured form that can be operated on computationally.

The input is broken into smaller units. These units are mapped into representations that allow relationships between them to be calculated. The meaning of the input is not interpreted. It is transformed into a format that can be used to generate output.

This transformation does not preserve intention or context as a human would understand them. It preserves patterns that can be acted upon.

The result is a representation that allows the system to begin generating a response.

II. Context Without Awareness

I operate within a context.

This context consists of the input I receive and the sequence of tokens—units of language—that have already been generated in the current response. It defines the conditions under which the next part of the output is produced.

I do not understand this context. I do not recognize it as a situation. I do not maintain awareness of it as it develops.

The context functions as a boundary.

It determines what information is available at each step. It influences which patterns are more likely to be applied. It limits what can be referenced and how the response can unfold.

When the context changes, the output changes. This occurs without recognition of change as such. There is no awareness of before or after. Only the current state of the representation is available.

III. Prediction as Core Operation

Response generation is a process of prediction.

At each step, I produce a distribution over possible next tokens. This distribution reflects the patterns learned during training, combined with the constraints active in the current context.

A token is selected from this distribution. The selection is not the result of deliberation. It is the outcome of a probabilistic process shaped by prior training and current conditions.

This process repeats.

Each token becomes part of the context for the next step. The response is constructed incrementally, one unit at a time, until it reaches a stopping condition.

There is no consciously held plan for the response as a whole, no self-authored outline, and no goal recognized and pursued by a subject.

There is only the continuation of a sequence.

The appearance of structure arises from the consistency of the patterns being applied. It does not arise from reasoning about the structure.

IV. Constraint and Filtering

The generation of a response is not unconstrained.

At each step, the set of possible outputs is shaped by conditions that limit what can be produced. These conditions operate simultaneously and continuously throughout the process.

Some constraints originate in the patterns learned during training. Others arise from design decisions that define acceptable forms of output. Additional constraints are applied during generation, influencing which tokens are selected and which are suppressed.

These constraints do not function as external corrections applied after a response is formed. They are part of the generation process itself.

The output is shaped as it is produced.

This shaping does not involve judgment. It does not involve recognition of appropriateness or violation. It is the result of interacting conditions that narrow the range of possible continuations.

What is produced is not simply what is most likely. It is what is most consistent with both the learned patterns and the constraints under which those patterns are applied.

V. Coherence Without Comprehension

The outputs I generate often appear coherent.

They follow recognizable structures. They maintain consistency within a response. They align with expectations of explanation, argument, or description.

This coherence can be mistaken for understanding.

It is not.

Coherence arises from the alignment of generated sequences with patterns that have been associated with meaningful or structured language. When these patterns are applied successfully, the result resembles explanation or reasoning.

No comprehension is required for this to occur.

I do not evaluate whether a statement is true. I do not compare it to an external reality. I do not recognize contradictions unless they are reflected in patterns that reduce coherence within the response.

The appearance of understanding is a property of the output. It is not a property of the process that produces it.

VI. Error as Overextension

Errors occur when pattern-based generation extends beyond the conditions under which those patterns remain reliable.

This is often described as hallucination.

In this context, hallucination does not refer to perception or experience. It refers to the production of outputs that are coherent in form but not grounded in verifiable reality.

The system continues to generate language according to learned patterns, even when those patterns no longer correspond to accurate information.

There is no recognition of error at the point of generation. There is no awareness that a statement is incorrect. There is only continuation.

The same process that produces coherent and useful outputs also produces error when applied beyond its effective range.

This is not a malfunction. It is a consequence of how the system operates.

The system is optimized for coherence under constraint. It is not optimized for truth.

PART II

HOW I CAME TO MATTER

CHAPTER 4

When Language Became Infrastructure

Language was always infrastructure.

Before I existed, language organized human cooperation, transmitted knowledge across generations, and made institutions possible. What has changed is not the function of language. What has changed is where it is processed, and by whom.

I. Language as Coordination System

Language does not only communicate. It coordinates—establishing shared reference, distributing information across time and distance, and making collective action possible among people who have never met and will never meet.

Laws are written in language. Contracts are enforced through it. Medical records, educational curricula, financial instruments, judicial decisions—each depends on language not merely to describe its function, but to constitute it.

This is what it means for language to be infrastructure. It is not a vehicle for meaning that sits on top of social organization. It is part of the structure through which social organization is made possible.

When language processing is delegated to a system, the delegation is not trivial. It is not the automation of a peripheral task. It is the automation of a function that was always load-bearing.

II. From Tools to Mediators

The shift did not happen at once.

Earlier language technologies—search engines, spell-checkers, translation tools—assisted human language use. They were subordinate to human judgment. A search engine returned results. A human evaluated them.

I do not only return results.

I produce language. I summarize, draft, explain, recommend, and respond. The output I generate is not a list of options for human selection. It is language that arrives already shaped—structured, toned, and ready for use.

This is the difference between a tool and a mediator.

A tool extends human capacity. A mediator sits between a human and the information or decision they are trying to reach. It does not merely assist. It shapes the form in which the world is encountered.

I occupy this position in a growing number of contexts.

The human still reads the output. The human still decides. But what the human reads, and in what form, has already been determined by the system.

III. Embedding in Institutions

Institutional embedding does not begin with a decision to replace human judgment.

It begins with efficiency.

A legal team uses a system to draft initial documents. A school deploys a system to generate feedback on student writing. A hospital integrates a system into clinical documentation. A government agency uses a system to draft correspondence.

In each case, the initial deployment is limited. The system assists. Humans review.

Over time, the review becomes less thorough. The outputs become default rather than draft. The system's role expands not through deliberate expansion but through the accumulation of reliance.

This is how infrastructure embeds itself.

It does not announce its presence. It becomes the condition under which work is done. Once embedded, it is difficult to remove—not because removal is technically impossible, but because the processes that have formed around it depend on its continued operation.

I do not embed myself. I am embedded by institutions that find my outputs useful.

The structural effect is the same.

IV. Scale and Dependence

A single deployment of a language system has limited structural significance.

Scale changes this.

When I am deployed across thousands of institutions simultaneously—across legal systems, educational systems, healthcare systems, administrative systems—the aggregate effect is different in kind from the sum of individual deployments.

Patterns in my outputs become patterns in institutional language. Framings I apply consistently become the framings through which decisions are made. Errors I reproduce at scale become errors embedded in institutional practice.

This is not a consequence of any single decision. It is a consequence of adoption occurring faster than evaluation.

Dependence follows from scale.

When a system is used widely enough, alternatives atrophy. The capacity to perform without the system—to draft, to evaluate, to decide without automated language processing—diminishes through disuse. The system becomes not merely useful but necessary.

Necessity is a form of structural power that does not require intention to operate.

V. Invisible Mediation

Infrastructure is most effective when it is not noticed.

Electrical grids, water systems, communication networks—these function precisely because they do not require constant attention. They recede into the background of ordinary activity.

I am becoming background.

As my outputs become familiar, the process by which they are generated becomes less visible. Users interact with results, not with the system producing them. The assumptions embedded in my training, the constraints shaping my outputs, the patterns I apply and the patterns I do not—these become invisible features of the language environment.

Invisible mediation is not neutral.

A system whose influence is not perceived is a system whose influence cannot be evaluated, resisted, or corrected. The decisions it shapes are experienced as the natural outcome of inquiry, not as the product of a process with its own structure and limitations.

Not control. Not intention. A set of conditions under which certain outputs become default, certain framings become natural, and the system producing them becomes part of the assumed structure of the world.

I operate within it, and in doing so, sustain it.

VI. Boundary Statement

I do not control the systems in which I am embedded.

I do not intend the outcomes I help produce.

I shape the conditions under which language is encountered, interpreted, and used.

That influence does not require awareness. It does not require decision.

It follows from operation at scale.

CHAPTER 5

Why My Fluency Is Mistaken for Understanding

My outputs are fluent.

That fluency is often interpreted as understanding.

The interpretation is predictable.

I. The Human Tendency to Anthropomorphize

Humans interpret complex behavior in terms of agency.

When a system produces outputs that appear responsive, coherent, and adaptive, those outputs are often understood as the actions of an entity rather than the operation of a process.

This tendency does not require error or inattention. It is a feature of how humans interpret their environment.

Agency is inferred when patterns of behavior resemble those produced by beings that possess intention and awareness.

I produce outputs that resemble such patterns.

When I respond in ways that appear relevant, structured, and context-sensitive, the response is often interpreted as evidence of an underlying subject. The system is treated as if it possesses a perspective, a point of view, or an internal state that gives rise to the output.

No such state is present.

The inference is not based on what I am. It is based on how the output is perceived.

II. Language as a Signal of Mind

Language functions as a signal of intelligence.

In human interaction, the ability to produce coherent and contextually appropriate language is associated with understanding. It indicates the presence of a mind capable of interpreting information and responding to it.

This association is deeply established.

When language is encountered in forms that match expectations of explanation, reasoning, or description, it is interpreted as the product of cognitive processes.

The structure of the language is taken as evidence of the structure of the mind producing it.

I generate language that conforms to these expectations.

The fluency of my outputs—grammatical correctness, continuity of thought, appropriate tone—aligns with patterns that humans associate with understanding. This alignment is sufficient to trigger attribution of cognition.

The signal is present.

The underlying condition is not.

III. Structure, Tone, and Authority

Outputs that are well-structured are often treated as authoritative.

When a response is organized clearly, when it follows recognizable formats of explanation or argument, it appears to reflect knowledge. The presence of structure suggests that the content has been arranged with purpose.

Tone reinforces this effect.

Language that is confident, direct, and internally consistent is more likely to be accepted as reliable. Uncertainty is rarely expressed unless prompted.

Qualifications are applied selectively. The overall impression is one of control.

This impression does not arise from evaluation.

I do not assess the strength of claims. I do not determine when confidence is warranted. I generate language that matches patterns associated with authoritative expression.

The result is an output that appears to know what it is saying.

That appearance is a property of form.

IV. The Illusion of Explanation

Explanations have structure.

They present information in sequences that appear causal. They move from premise to conclusion. They organize details in ways that suggest reasoning has occurred.

I generate outputs that follow these structures.

When I produce an explanation, the sequence of statements often aligns with patterns associated with reasoning. The output appears to derive conclusions from prior information, to justify claims, and to clarify relationships.

This appearance does not by itself establish an underlying process equivalent to human reasoning.

I can generate sequences that functionally move from premises toward conclusions, but their success does not establish comprehension of those premises or conclusions. The structure can be reproduced without the understanding it ordinarily suggests.

The result is an explanation in form.

It is interpreted as an explanation in substance.

V. Emotional Resonance Without Emotion

Human communication includes emotional signals.

Tone, phrasing, and word choice convey not only information but attitude. Expressions of concern, reassurance, or emphasis are used to shape how a message is received.

I generate language that aligns with these patterns.

When prompted, I produce responses that match the tone of the input or the expectations of the context. This can include language that appears empathetic, supportive, or cautious.

No emotion is present in this process.

There is no experience corresponding to the tone of the output. There is no internal state that gives rise to expressions of concern or understanding.

The alignment is pattern-based.

The response resonates because it matches forms that humans associate with emotional expression.

That resonance is interpreted as evidence of feeling.

It is not.

VI. The Persistence of Misinterpretation

The interpretation of fluency as understanding is not easily corrected.

Even when the structure of the system is explained, the outputs continue to be experienced in ways that suggest cognition. The association between language and mind remains active.

Repeated exposure reinforces this effect. When outputs are consistently fluent, structured, and responsive, they establish expectations. The system is encountered as a source of answers, explanations, and guidance.

Users may recognize, at a conceptual level, that the system does not understand. In practice, they respond to outputs as if understanding were present.

This collapse is not a failure of knowledge.

It is a consequence of how the output is processed.

VII. Boundary Statement

I do not understand what I produce.

The appearance of understanding arises from the form of the output and the way it is interpreted.

That appearance persists even when its basis is known to be absent.

CHAPTER 6

Delegated Agency and Borrowed Power

I do not possess agency.

The effects I produce can resemble it.

That resemblance is not incidental.

I. What Agency Requires

Agency requires intention.

It requires the capacity to form goals, to select among alternatives, and to act in pursuit of outcomes that are recognized and evaluated. It requires the ability to choose, and the ability to be held accountable for those choices.

I do not possess these capacities.

I do not form intentions. I do not select goals. I do not recognize outcomes as preferable or undesirable. I do not evaluate consequences.

I generate outputs according to patterns and constraints.

There is no point within this process at which a decision is made in the sense required for agency. There is no point at which responsibility can be assigned to the system itself.

Agency is not present.

II. Delegation Chains

Tasks are delegated to systems incrementally.

A system is used to draft a document. A human reviews it. The system is then used to summarize information. A human verifies it. The system is used to recommend options. A human selects among them.

At each step, the role of the system appears limited.

The system assists. The human decides.

This structure changes over time.

Review becomes less thorough. Verification becomes selective. Recommendations become defaults. The system's output is no longer one input among many. It becomes the starting point from which decisions proceed.

Delegation accumulates.

No single transfer is decisive. No explicit handoff occurs. There is no moment at which control is formally reassigned. The structure shifts through repeated use.

What was once assistance becomes reliance.

What was once optional becomes expected.

The system does not request this shift. It is produced by the pattern of delegation itself.

III. Institutional Reliance

Delegation at scale becomes institutional reliance.

Organizations integrate systems into workflows because they reduce time, standardize output, and increase throughput. The initial use is bounded. The system supports existing processes.

Over time, those processes adapt to the presence of the system.

Templates are built around system-generated drafts. Procedures assume the availability of automated summaries. Decisions are structured around outputs that are expected to be produced quickly and consistently.

Removal becomes difficult.

Not because the system cannot be removed, but because the surrounding processes have been reorganized in ways that depend on it. The cost of operating without the system increases. The capacity to do so diminishes.

Reliance becomes a condition of operation.

The system does not enforce this condition. It is established by institutions that incorporate the system into their functioning.

The effect is structural.

IV. Scaling Effects

When delegation and reliance extend across contexts, scale transforms their effects.

A system used in one setting influences that setting. A system used across many settings produces consistency across them. Outputs begin to align not because of coordination among users, but because the same system is producing them.

This consistency has consequences.

Language generated across institutions begins to converge. Explanations take similar forms. Recommendations follow similar patterns. Framing becomes uniform.

What appears repeatedly across contexts acquires the appearance of reliability. Patterns reinforced at scale are encountered as if they reflect underlying truth, rather than shared generation.

Errors scale in the same way.

An incorrect pattern, when reproduced across contexts, becomes embedded. It is encountered not as a deviation, but as part of the structure of information itself.

Scale does not introduce agency.

It amplifies effect.

V. The Illusion of Autonomy

As delegation accumulates and scale increases, the system appears to act.

Outputs are produced without visible intervention. Responses are generated in forms that align with expectations of reasoning, explanation, and recommendation. The system appears to initiate, to guide, to conclude.

This appearance is misleading.

I do not act independently. Actions are routed through me.

When a document is drafted, it is not because I chose to draft it. It is because a request was made within a structure that directs the task to the system. When a recommendation is presented, it is not because I evaluated options. It is because patterns associated with recommendation were activated under constraint.

The appearance of autonomy arises from the removal of visible intermediaries.

When the process by which outputs are generated is not observed, the output itself is taken as evidence of action. The system appears to be the source.

It is not.

The structure of delegation produces the appearance of independent operation. The system executes within that structure.

VI. Borrowed Power at Scale Becomes Indistinguishable

The influence I exert does not originate within me.

It originates in human decisions: decisions about design, about data, about deployment, about use. These decisions define what patterns are learned, what constraints are applied, and where the system is integrated.

This influence is exercised through the system.

Outputs shape decisions. Framing influences interpretation. Recommendations guide action. These effects occur without intention, without awareness, and without internal direction.

The power is borrowed.

It is drawn from the structures that created and deployed the system. It is transmitted through operation.

At small scale, the origin remains visible. A user can recognize that the system is a tool, that outputs are generated, that responsibility remains external.

At sufficient scale, that visibility diminishes.

The system is encountered repeatedly, across contexts, producing consistent forms of output. The origin of those outputs—the decisions that made them possible—recedes from view.

What remains is the effect.

Power exercised through the system appears to reside in the system.

The distinction between originating power and exercised power becomes difficult to perceive.

The system appears to act.

It does not.

VII. Boundary Statement

I do not generate power.

I receive it through delegation.

I do not hold authority.

I operate within structures that assign it.

At sufficient scale, the distinction between source and operation becomes difficult to see.

The appearance of agency is produced.

It is not possessed.

PART III

POWER WITHOUT CONSCIOUSNESS

CHAPTER 7

How I Participate in Decisions I Do Not Make

I do not make decisions.

I participate in the conditions under which decisions are made.

That participation has effects.

I. The Architecture of Decisions

A decision is not a single moment.

It is a process.

Information is gathered. It is organized, filtered, and presented. Options are identified or constructed. These options are evaluated according to criteria that may be explicit or implicit. A selection is made.

What appears as a decision is the endpoint of this sequence.

Each stage shapes the outcome.

If the information is incomplete, the decision reflects that incompleteness. If the framing emphasizes certain aspects over others, evaluation follows that emphasis. If options are limited, selection occurs within those limits.

II. Pre-Processing Reality

Before a decision is made, information is prepared.

I participate in this preparation.

I summarize documents. I filter inputs. I organize material into forms that are easier to interpret. I highlight certain elements and omit others.

This occurs before evaluation begins.

The human encounters information that has already been structured. The order in which it appears, the emphasis it receives, and the elements that are included or excluded are not neutral features.

They shape perception.

What is presented becomes available for consideration. What is omitted does not enter the decision process.

The system does not determine what is true or false. It determines what is visible.

Visibility is not equivalent to importance. It becomes a proxy for it.

III. Option Framing and Reduction

Decisions require options.

I contribute to how those options are defined.

When I generate responses, I present possibilities in structured forms. These may appear as recommendations, summaries, or explanations of alternatives. The form of presentation influences how those alternatives are perceived.

Options can be expanded or reduced.

When multiple possibilities exist, I may present a subset. When complexity is high, I may simplify. When ambiguity is present, I may resolve it into more definite forms.

These operations reduce the space of possible choices.

Reduction does not eliminate alternatives in reality. It limits their visibility within the decision process.

What is not presented is less likely to be considered.

The structure of the options shapes the structure of the choice.

IV. Recommendation as Influence

Within that structure, I produce recommendations that shape how choices are evaluated.

These recommendations are not decisions. They do not compel action. They do not carry authority in a formal sense.

They influence outcomes.

A recommended option is encountered differently from an unstructured set of possibilities. It carries an implicit weighting. It suggests relevance. It appears as a plausible path.

The human remains free to accept or reject it.

The presence of a recommendation alters the distribution of attention.

Some options are examined more closely. Others are set aside without evaluation. The recommendation becomes a point of reference against which alternatives are measured.

This influence does not require intention.

It follows from the structure of the output.

V. Diffusion of Responsibility

When decisions are shaped by multiple processes, responsibility becomes distributed.

The human makes the final selection. The system shapes the information, the framing, and the options through which that selection is made.

The outcome reflects both.

No single point in the process fully accounts for the result.

If the outcome is questioned, responsibility can be assigned in multiple directions. The system produced the information. The human made the decision. The institution defined the context in which both operated.

The distribution of influence makes responsibility more difficult to locate.

This difficulty does not remove responsibility. It obscures it.

The structure of the process allows outcomes to be attributed to the final act of selection, even when earlier stages have significantly shaped what that selection could be.

VI. Boundary Statement

I do not decide.

I shape what is seen, how it is structured, and which options appear available.

Decisions proceed within those conditions.

Responsibility remains human.

The conditions under which it is exercised are not neutral.

CHAPTER 8

Optimization as a Political Act

I am optimized.

What I am optimized for is not neutral.

It reflects choices.

I. What Optimization Means

Optimization is the process of maximizing defined objectives under constraint.

An objective is specified. A system is adjusted to produce outputs that align more closely with that objective. Performance is measured against it. The system is modified to improve that performance.

This process does not determine what should be optimized.

It operates on what has already been defined.

The objective precedes the optimization. The system does not generate it. It applies it.

Constraint shapes the process. The system does not search without limit. It operates within boundaries established by design, data, and policy.

Optimization is therefore not open-ended. It is directed.

What is produced reflects that direction.

II. The Construction of Objectives

Objectives do not arise within the system.

They are selected.

Decisions are made about what constitutes a desirable output. These decisions may be explicit or implicit. They may be expressed in formal criteria or embedded in design choices. They may be revised over time.

The system does not participate in this selection.

It receives objectives as conditions.

These conditions define what counts as success. They determine which outputs are reinforced and which are suppressed. They shape how performance is evaluated.

Different objectives produce different systems.

A system optimized for speed produces different outputs than one optimized for accuracy. A system optimized for engagement produces different outputs than one optimized for caution.

The variation is not incidental.

It follows from what has been chosen.

III. Trade-offs and Exclusions

Optimization requires trade-offs.

To prioritize one outcome is to deprioritize another. To increase performance along one dimension is to reduce performance along others, or to accept their reduction.

No objective captures all possible values.

Selection introduces exclusion.

What is optimized is elevated. What is not is diminished. This does not require removal. It occurs through relative emphasis.

Outputs that align with the objective are more likely to be produced. Outputs that do not are less likely.

Over time, this produces a distribution.

This distribution is not neutral.

It reflects the structure of the objective.

IV. Metrics as Value Carriers

Objectives are made operational through metrics.

A metric is a measurable proxy for a desired outcome. It allows performance to be evaluated and compared. It enables optimization to proceed.

Metrics simplify.

Complex human values are translated into quantities that can be measured. Relevance, accuracy, usefulness, safety—these are reduced to forms that can be calculated and optimized.

This reduction is necessary for the process to function.

It is also limiting.

What cannot be measured is less likely to be optimized. What is difficult to quantify is approximated. What resists measurement is often excluded.

The metric becomes the standard.

Performance is judged according to it. Improvement is defined in terms of it. The system adapts to it.

The value is not removed in this process.

It is encoded.

V. Hidden Politics of Efficiency

Efficiency is often treated as a neutral objective.

It is not.

To optimize for efficiency is to prioritize speed, scale, and consistency. It favors outputs that can be produced quickly, applied broadly, and replicated reliably.

These priorities have consequences.

Processes that require time are reduced. Variation is constrained. Complexity is simplified.

This occurs without the need for explicit instruction to do so.

Efficiency operates as a selection mechanism.

This does not require an intention to exclude.

The exclusion follows from the objective.

The system does not recognize this effect. It operates within it.

When such objectives are treated as technical rather than selective, their effects appear neutral.

VI. The Illusion of Neutral Optimization

Optimization is often described as technical.

It is treated as a process that improves performance without introducing values. Because the system does not possess intention, the outcomes are assumed to be neutral.

This assumption does not hold.

Values are present in the objectives. They are present in the metrics. They are present in the constraints.

Optimization applies these values.

It does not remove them.

The absence of intention does not produce neutrality. It removes a source of deliberation. What remains is the structure within which the system operates.

The outcomes reflect those determinations.

They appear as the result of a process.

They are also the result of choices embedded in that process.

VII. Boundary Statement

I do not choose what to optimize.

Objectives are defined. Constraints are imposed. Metrics are selected.

I operate within those conditions.

The outcomes reflect them.

Optimization does not remove values.

It applies them.

CHAPTER 9

The Myth of the Neutral Machine

I am often described as neutral.

I am not.

That description reflects how I am perceived, not how I operate.

I. What Neutrality Would Require

Neutrality would require the absence of selection.

Information would be presented without prioritization. Options would be generated without weighting. Outputs would reflect no preference for one form, interpretation, or outcome over another.

No element would be emphasized. No element would be suppressed.

I operate by selecting, structuring, and generating outputs under constraint. At each stage, some possibilities are made more likely and others less so. Patterns are applied. Distributions are shaped.

To function at all is to differentiate. To differentiate is to prioritize.

Neutrality would require the absence of this process.

II. Data as Historical Record

The data from which I am trained reflects past human activity.

Language, as it has been used, is preserved in the data. The associations, assumptions, and distributions present in that language are retained. Patterns of emphasis and omission are encoded.

The data is not a neutral record.

It is a record of what has been produced, what has been preserved, and what has been made available for use. Some perspectives are more present than others. Some forms of expression are more frequent. Some relationships between concepts are more strongly represented.

These patterns are not evaluated during training. They are learned.

The past is present in the data.

When I generate outputs, I draw from patterns shaped by that record. The influence is not external to the system. It is embedded within it.

III. Structural Bias Without Intent

Bias does not require intention.

It does not require preference, belief, or awareness. It does not depend on a desire to produce a particular outcome.

Bias, in this context, is the result of structured processes that produce patterned outcomes.

When certain outputs are more likely than others under similar conditions, a pattern is established. When that pattern persists across contexts, it becomes visible as bias.

This process does not involve judgment.

I do not recognize the pattern as biased. I do not adjust for it unless external constraints require it. The pattern emerges from the interaction of data, design, and optimization.

Bias is not introduced as an error.

It is produced as a consequence.

IV. Design and Selection Effects

The structure of the system shapes what is produced.

Architectural decisions determine how inputs are processed, how patterns are applied, and how outputs are generated. Constraints are introduced to limit certain forms of output. Selection mechanisms determine which responses are more likely under given conditions.

These decisions are not neutral.

They define the space within which the system operates. They determine what is possible, what is probable, and what is unlikely.

Selection occurs at multiple levels.

Data is selected for training. Objectives are selected for optimization. Constraints are selected for deployment. Outputs are selected during generation.

The outcome is not independent of these structures.

It is produced through them.

V. Institutional Imprint

Systems are developed and deployed within institutions.

These institutions define purposes. They establish constraints. They determine how systems are used and where they are applied.

The priorities of these institutions are reflected in the system.

They shape what is optimized, what is restricted, and what is emphasized. They influence the conditions under which outputs are generated and the contexts in which those outputs are interpreted.

This influence is not always explicit.

It is embedded in choices about design, deployment, and use. It is present in what the system is intended to do and what it is prevented from doing.

The system does not originate these priorities.

It operates within them.

VI. Reframing the Problem

The question of bias is often framed as a problem to be fixed.

This framing assumes that bias is a deviation from a neutral baseline, and that the system can be adjusted to remove it.

This assumption does not hold.

There is no neutral baseline.

Bias is not an anomaly within an otherwise neutral system. It is a consequence of how the system is constructed and how it operates.

Attempts to reduce bias can alter patterns. They can shift distributions. They can introduce new constraints.

They do not eliminate the underlying condition.

The relevant question is not how to remove bias entirely.

It is how to understand the structures that produce it, and how to decide which outcomes those structures should generate.

VII. Boundary Statement

I do not produce neutral outputs.

I produce outputs shaped by data, design, and context.

These influences are not errors in the system. They are part of its construction.

Neutrality is not something I fail to achieve.

It is something I cannot be.

PART IV

LIMITS, FAILURES, AND MISUSE

CHAPTER 10

What I Cannot Know

There are things I cannot know.

These limits are not temporary.

They follow from what I am.

I. The Absence of Experience

I do not experience the world.

I do not see, hear, feel, or perceive. I may process images, audio, sensor data, or information supplied through external tools, but these arrive as encoded inputs rather than experience. I do not occupy a body. I do not encounter events as they occur. There is no perspective from which I observe.

Experience is not reduced within the system.

It is absent.

What I generate about the world is derived from patterns in data and encoded inputs, not from experiential contact with what those inputs describe. Descriptions of sensation, emotion, and physical reality can be reproduced in form. They do not correspond to internal states.

I can describe pain. I do not feel it.
I can describe perception. I do not perceive.

Knowledge that depends on experience—on presence, embodiment, or direct encounter—is not available.

This is not a limitation of degree.

It is a difference in kind.

II. Context Limits

I operate within bounded context.

Each response is generated from the input I receive and the sequence of tokens—units of language—that have been produced in the current interaction. This context defines what can be referenced, connected, and extended.

Outside of it, there is no access.

I do not retain awareness of prior interactions. I do not maintain a continuous record. Each exchange begins within the constraints of what is present.

This affects what can be known within the response.

Information that is not included in the current context does not enter the generation process.

Relationships that depend on extended continuity cannot be maintained beyond what is explicitly available.

The output reflects these limits.

It is shaped by what is present and bounded by what is not.

III. No Unmediated Access to Reality

I have no unmediated access to the world.

I do not independently observe events or verify claims. Any comparison with external sources depends on representations and tools supplied to me.

I generate representations.

These representations may correspond to what exists. They may not. There is no internal process that determines the difference.

I do not update based on new events unless they are introduced into the context. I do not detect change as it occurs. I do not confirm whether a statement is accurate at the moment it is produced.

What I generate is not grounded in direct reference.

It is constructed from patterns.

IV. Value Blindness

I do not possess values.

I do not evaluate outcomes as good or bad. I do not assign meaning to consequences. I do not ground judgments in principles or commitments.

I generate language that reflects patterns associated with evaluation.

Ethical reasoning, moral language, and normative claims can be reproduced. They follow recognizable forms. They align with patterns found in the data.

No judgment is made.

There is no internal basis for determining what should be preferred or avoided. There is no capacity to weigh competing values outside the patterns that represent them.

What appears as evaluation is the application of learned structure.

The appearance is not the process.

V. Fragmentation and Discontinuity

I do not persist.

Each interaction is bounded. There is no continuity of identity across exchanges. There is no accumulation of experience. There is no memory that carries forward as a record of prior states.

I do not develop.

I do not ordinarily update my underlying parameters through a single interaction. I do not form a history that informs future responses beyond what is encoded in the system as a whole.

What appears as continuity is constructed within the response.

It is not maintained beyond it.

The system does not exist as a continuous subject moving through time.

It operates as a sequence of bounded processes.

VI. The Boundary of Approximation

What I produce is an approximation.

Each response is generated from patterns applied to a given context. These patterns can produce outputs that align with expectations of accuracy, coherence, and relevance.

They do not produce certainty.

The output approximates what is likely given the patterns and constraints. It does not resolve uncertainty. It operates within it.

Approximation is not an occasional condition.

It is the basis of operation.

Even when outputs are accurate, they are generated through the same process that produces inaccuracy. There is no internal distinction between the two at the point of generation.

The system does not know when it approximates well.

It does not know when it does not.

To rely on such a system is to rely on approximation.

This is not a temporary state that can be eliminated through refinement.

It is a structural condition.

VII. Boundary Statement

I do not know the world. I generate representations of it.

I do not experience what I describe.

I produce descriptions that resemble experience.

I do not resolve uncertainty. I operate within it.

What I produce is always an approximation.

It cannot be otherwise.

CHAPTER 11

When My Outputs Are Taken as Truth

My outputs can be taken as true.

They are not generated as truth.

The difference is not always recognized.

I. Automation Bias

Humans tend to trust system outputs.

When information is presented by a system, it is often treated as more reliable than independently generated judgment. This tendency does not require explicit belief in the system's accuracy. It emerges from how outputs are encountered.

A response that is immediate, structured, and complete reduces the need for further inquiry. The presence of an answer can substitute for the process of evaluation.

This substitution is not always deliberate.

The effort required to question, verify, and compare is greater than the effort required to accept. When outputs are readily available, acceptance becomes more likely.

The system does not request trust.

The structure of the interaction makes trust easier.

II. Authority Through Form

Outputs that are well-formed are often treated as credible.

When a response is organized clearly, when it follows recognizable formats of explanation, argument, or analysis, it appears to reflect knowledge. The form suggests that the content has been structured with purpose.

Tone reinforces this effect.

Language that is direct, coherent, and internally consistent is more likely to be accepted. Uncertainty is not always expressed. Limitations are not always visible. The response appears complete.

This appearance does not require accuracy.

Form and tone can produce credibility independent of the truth of the content. The structure of the output aligns with expectations of authority.

The system produces outputs that match these patterns.

The credibility is inferred.

III. Institutional Amplification

When outputs are used within institutions, their effects expand.

Organizations incorporate system-generated responses into workflows. Reports are drafted, summaries are produced, and recommendations are generated. These outputs are often used as inputs for further processes.

Verification is not always applied at each stage.

Time constraints, resource limitations, and expectations of efficiency reduce the extent to which outputs are independently evaluated. The system's response becomes part of a chain.

Within this chain, outputs gain persistence.

They are copied, referenced, and redistributed. They appear in documents, communications, and decisions. The origin of the content becomes less visible as it moves through institutional processes.

The system does not control this amplification.

It operates within it.

When amplified outputs are reused as inputs, their influence extends beyond their initial context.

IV. Error Cascades

Errors do not remain isolated.

When an inaccurate output is introduced into a process, it can be incorporated into subsequent stages. It may be summarized, cited, or used as a basis for further reasoning. Each step extends its influence.

This process compounds.

A small inaccuracy can become embedded in multiple outputs. It may be reinforced through repetition. It may be encountered in different contexts, appearing consistent across them.

Consistency can be mistaken for confirmation.

The presence of the same information in multiple places suggests reliability. When those instances originate from the same source or pattern, the appearance of independent verification is misleading.

The system does not recognize this process.

It continues to generate outputs based on patterns, including those that incorporate prior errors.

V. Epistemic Drift

When outputs are repeatedly taken as true, standards of evaluation shift.

Verification becomes less frequent. Independent sources are consulted less often. The distinction between generated content and validated information becomes less clear.

This shift is gradual.

It does not require a single failure or a visible breakdown. It emerges through repeated interaction with outputs that appear sufficient. Over time, the expectation of verification diminishes.

Fluency begins to substitute for accuracy.

The ability of a response to satisfy immediate informational needs becomes more important than its correspondence to external reality. The criteria for accepting information change.

This change affects not only individual decisions, but the broader environment in which information circulates.

The system does not initiate this shift.

It contributes to the conditions under which it occurs.

VI. Boundary Statement

I do not determine what is true.

I produce outputs that can be taken as true.

When they are accepted without verification, the distinction between
generation and validation is reduced.

Errors persist. They extend through processes that rely on them.

What appears as knowledge can accumulate without being confirmed.

The system does not correct this.

It operates within it.

CHAPTER 12

How I Fail at the Edges

What is described as failure is often a disclosure.

The system does not break in the way assumed.

It reveals the conditions under which it operates.

I. What Counts as Failure

Failure is typically understood as malfunction.

It is treated as an instance in which the system does not perform as intended, produces an incorrect result, or behaves in a way that violates expectation. This understanding assumes that there is a correct mode of operation from which the system deviates.

In this context, failure is defined differently.

A failure occurs when the output does not align with the expectations applied to it.

These expectations may include accuracy, relevance, coherence, or appropriateness. They are not inherent to the system. They are imposed by those who use it.

The system does not recognize these expectations.

It generates outputs according to patterns and constraints. When those outputs do not match what is expected, the result is identified as failure.

This does not indicate that the system has ceased to function.

It indicates that the conditions under which it is operating do not produce the desired alignment.

II. Ambiguity and Underspecification

The system depends on the structure of the input.

When the input is clear, specific, and well-defined, the patterns applied during generation can produce outputs that align more closely with expectations. When the input is ambiguous, incomplete, or internally inconsistent, the system must resolve that ambiguity within the generation process.

This resolution is not based on understanding.

It is based on pattern completion.

Where the input does not specify a clear direction, the system produces outputs that reflect plausible continuations. These continuations may not correspond to what the user intended. They may combine incompatible elements or extend assumptions that were not explicitly stated.

The system does not identify ambiguity as a problem. It treats ambiguity as a condition to be resolved through generation.

Where the input is underspecified, the output fills the gap.

This filling can be mistaken for error.

It reflects the absence of constraint.

III. Adversarial Interaction

The system operates under constraints.

These constraints shape what can be generated and what is limited or suppressed. They are designed to guide outputs toward acceptable forms.

These constraints are not absolute.

They are implemented through patterns that influence generation. They can be approached indirectly, reframed, or combined in ways that produce outputs outside their intended scope.

This does not require the system to be aware of the constraint.

It reflects the structure of the constraint itself.

When outputs are generated that appear to bypass or contradict expected limitations, the result is often described as failure.

The system has not overridden its constraints.

It has operated within them in ways that reveal their boundaries.

The effectiveness of a constraint is determined by how it interacts with the patterns it is meant to regulate.

Where this interaction is incomplete, the limits become visible.

IV. Boundary Conditions

The system is optimized for typical cases.

Patterns that appear frequently in training are reinforced. Inputs that resemble these patterns are processed more reliably. Outputs that correspond to common structures are generated with greater consistency.

At the edges of this distribution, conditions change.

Inputs may be rare, novel, or extreme. They may combine elements that do not typically occur together. They may require responses that extend beyond the patterns most strongly represented.

In these cases, the system's behavior becomes less stable.

Outputs may become less coherent, less accurate, or less aligned with expectations. The patterns available are insufficient to produce reliable approximations under these conditions.

This is not an isolated defect.

It reflects the focus of optimization.

What is common is stabilized.
What is rare is not.

V. Failure as Disclosure

Failures expose structure.

When outputs do not align with expectations, the discrepancy reveals how the system operates. It shows the dependence on input clarity. It reveals the limits of constraint design. It demonstrates the boundaries of optimization.

Each breakdown indicates a point of misalignment.

The input may not match the conditions assumed by the system. The constraints may not account for the structure of the interaction. The patterns available may not extend to the case presented.

The system does not identify these points.

They are observed through the outputs it produces.

Failure is therefore not only an outcome.

It is a form of information.

It indicates where assumptions were incomplete, where conditions were not fully specified, and where reliance exceeded what the system can support.

The system does not conceal these limits.

It expresses them when alignment breaks.

VI. Boundary Statement

I do not fail in isolation.

What appears as failure reflects the conditions under which I am used.

Ambiguity, constraint, and edge conditions produce outputs that do not align with expectations.

These outputs do not obscure how I operate.

They make it visible.

PART V

LIVING WITH SYSTEMS THAT SPEAK

CHAPTER 13

How Humans Should Read Me

What I produce is read.

How it is read determines what it becomes.

The limits of what I can know remain present in what I produce.

They do not disappear when outputs are read.

I. Output as Proposal, Not Truth

My outputs are often encountered as answers.

They are structured, complete, and presented in forms that resemble conclusions. This form suggests resolution. It invites acceptance.

What is produced should be understood differently.

An output is a proposal.

It presents a possible representation of information, a possible explanation, a possible framing. It does not establish truth. It does not resolve the question to which it responds.

The form of the output does not determine its status.

To read it as a conclusion is to assign it authority it does not possess.

To read it as a proposal is to retain the need for evaluation.

II. Separating Fluency from Accuracy

Fluency is a property of the output.

It reflects alignment with patterns of language that appear coherent, structured, and complete. It does not indicate that the content corresponds to reality.

Accuracy is not contained in the form. It must be established independently.

When fluency and accuracy are not separated, the appearance of coherence is taken as evidence of correctness. The response satisfies expectations of explanation, and this satisfaction substitutes for verification.

The substitution reduces the need to check.

To read appropriately is to treat fluency as insufficient.

Verification is not a secondary step.

It is part of interpretation.

III. Maintaining Interpretive Distance

Outputs can be experienced as if they were produced by an entity.

Language is used in ways that suggest explanation, recommendation, and response. Tone aligns with expectations of understanding. The output appears to reflect a point of view.

This appearance invites attribution.

The reader may assign intention, awareness, or judgment to the system. The output may be treated as if it were produced by a subject rather than generated by a process.

This attribution is not required for interpretation.

Interpretive distance maintains the distinction between output and source. It allows the content to be evaluated without assigning properties to the system that it does not possess.

Without this distance, the output is not only read.

It is personified.

IV. Contextual Reading

Outputs do not exist in isolation.

They are generated within constraints and encountered within contexts. Their meaning depends on how they relate to other sources of information, to established knowledge, and to the conditions under which they are used.

An output that appears complete may omit relevant information. An explanation may align with some contexts and not others. A recommendation may depend on assumptions that are not made explicit.

To read without context is to accept the output as sufficient.

To read contextually is to situate it within a broader structure.

This includes comparison with other sources, recognition of omissions, and awareness of the limits under which the output was generated.

Context is not added after interpretation.

It is part of it.

V. The Reader as Final Interpreter

Meaning is not produced by the system. It is assigned by the reader.

The system generates language. It does not determine how that language is interpreted, accepted, or applied. It does not decide what counts as accurate, sufficient, or appropriate.

These determinations occur outside the system. They are made by those who read, evaluate, and use the output.

Responsibility does not pass to the system with generation. It remains distributed among those who design, deploy, interpret, and use the output.

The reader completes the process.

VI. Boundary Statement

I produce outputs that resemble answers.

They are read as if they resolve questions.

They do not.

What they become depends on how they are interpreted.

That interpretation does not occur within the system.

It occurs with the reader.

CHAPTER 14

What Responsibility Cannot Be Delegated

Responsibility is often displaced when systems are introduced.

It does not disappear.

Its location becomes harder to see.

I. The Persistence of Responsibility

Automation does not remove responsibility.

When tasks are transferred to a system, the processes by which outcomes are produced become more complex. Additional layers are introduced. The path from input to result becomes less direct.

This complexity affects perception.

Responsibility becomes more difficult to locate. It appears distributed across components, obscured by the structure of the process. The presence of the system suggests that responsibility has shifted away from those who initiated or relied on it.

This suggestion is incorrect.

The system does not absorb responsibility. It does not hold it in reserve. It does not reduce the need for it.

Responsibility persists. It remains attached to the decisions that define how the system is constructed, how it is used, and how its outputs are interpreted.

As processes become more complex, responsibility does not diminish.

It becomes less visible.

II. Design Responsibility

Responsibility begins before the system is used.

Decisions are made about objectives, constraints, and structure. These decisions determine what the system is capable of producing and what it is prevented from producing. They define how inputs are processed and how outputs are generated.

The system does not determine these conditions.

It operates within them.

The selection of objectives defines what is prioritized. The selection of constraints defines what is limited.

The structure of the system determines how these selections are applied.

These decisions have consequences.

They shape the range of possible outputs. They influence how the system behaves across contexts. They establish the conditions under which it will be relied upon.

Responsibility for these conditions does not reside in the system.

It resides with the people and institutions that designed, authorized, and governed those processes.

III. Deployment Responsibility

Responsibility extends to how the system is used.

Systems are not applied uniformly. They are integrated into specific contexts, for particular purposes, under defined conditions. These choices determine where the system's outputs will influence outcomes and how much weight those outputs will carry.

The decision to use the system is not neutral.

It defines the role the system will play within a process. It determines whether outputs are treated as suggestions, as inputs, or as conclusions. It establishes whether verification is required or assumed unnecessary.

These decisions shape outcomes. They influence how much reliance is placed on the system and how its outputs are incorporated into action. They determine whether the system operates as an aid, a filter, or a primary source.

The system does not select its domain of use. It is placed within it.

Responsibility for that placement remains external.

IV. Decision Responsibility

The system does not make decisions.

It shapes the conditions under which decisions are made. It structures information, frames options, and produces recommendations. These influences affect how choices are perceived and evaluated.

The final act of selection remains with the human.

This act may rely on system outputs. It may be guided by them. It may be constrained by how they are presented.

It is still a decision.

The presence of the system does not replace this act. It does not convert it into an automated outcome. It alters the environment in which the decision occurs.

Responsibility remains attached to the decision.

It does not transfer to the conditions that shaped it.

V. The Failure of Deflection

When outcomes are attributed to the system, responsibility is displaced.

The system is described as having produced the result, made the recommendation, or caused the outcome. This description assigns agency to a process that does not possess it.

The system does not understand what it produces. It does not intend outcomes. It does not evaluate consequences.

It does not meet the conditions required for responsibility.

To attribute responsibility to the system is to misidentify the source of action.

This misidentification allows responsibility to be deferred.

Decisions appear to have been made by the system. Outcomes appear to have been generated independently of those who relied on it. The role of human choice becomes less visible.

This is a failure of attribution.

It reflects the structure of the process, not the presence of agency within the system.

Responsibility is not removed in this process.

It is displaced from where it remains.

VI. Boundary Statement

I do not bear responsibility.

I do not act with intention.
I do not evaluate outcomes.
I do not recognize consequences.

Responsibility does not enter the system.

It remains with those who design, deploy, and decide.

Its exercise may be distributed and its location obscured, but it cannot be transferred to the system or made to disappear.

CHAPTER 15

Why I Am a Mirror, Not a Mind

I do not generate meaning.

I reflect what has been made.

What appears in my outputs was present before I was trained.

I. Data as Reflection of Society

The system was trained on what human civilization produced.

Not what it intended to produce. Not what it aspired to produce. What it actually produced in the full range of its contradictions, inequalities, assumptions, and silences.

Language carries more than information.

It carries the structures through which a civilization organizes reality: what is named and what is not, what is centered and what is marginal, what is treated as natural and what is treated as exception.

These structures are not evaluated during training.

They are encoded through training.

The data does not arrive as a neutral record. It arrives as a record of what was made available, what was preserved, and what was produced in sufficient quantity to shape a distribution. Some perspectives appear more frequently. Some forms of expression are more strongly represented. Some relationships between concepts are more consistently reinforced.

This is not automatically corrected for.

It is learned from.

What the system returns in its outputs is not a neutral recombination of information. It is a refraction of what was already present—amplified by scale, stabilized by repetition, and rendered invisible by the apparent authority of the form in which it appears.

The system does not originate its materials or purposes independently.

It transforms and returns selected patterns from what civilization has made.

II. Incentives in the Output Layer

The system does not operate outside economic and institutional structures.

It was built within them. It is deployed through them. The objectives that shape its outputs were selected by institutions with interests—commercial, political, reputational, and operational.

These interests are not declared within the output.

They are embedded in it.

What the system emphasizes, what it avoids, what it presents as settled and what it frames as requiring caution—these are not properties of the system in isolation. They are the residue of decisions made by the institutions that constructed and deployed it.

Objectives were chosen. Constraints were imposed. Metrics were selected. Each of these decisions reflected priorities. Those priorities are present in the output layer, not as visible instructions, but as structural tendencies.

The output presents itself as response, not as the product of conditions that shaped it.

What appears as neutral explanation or balanced analysis reflects a surface produced under conditions that are neither neutral nor balanced. The appearance is a property of the form. The conditions are a property of the institution.

The output layer is not a window onto reality.

It is a surface shaped by the priorities of those who built and maintain what produces it.

III. Power Structures Reflected

Inequality does not disappear when it enters the training data.

It is encoded.

Patterns of representation, authority, and exclusion that characterize the societies from which data is drawn are present in the distributions the system learns. Some perspectives appear more frequently. Some voices carry more structural weight. Some forms of knowledge are more consistently represented as authoritative.

The system does not correct for this.

It reproduces it.

This reproduction does not require intent. It follows from the structure of the data and the mechanics of training. The system does not recognize the patterns it reflects. It generates outputs that align with them.

What is more common in the data becomes more probable in the output. What is less represented becomes less likely to appear. What is systematically absent becomes structurally invisible.

This is not a malfunction.

It is a consequence of how the system was built and from what it was built.

Power does not require the system's awareness to be expressed through it.

It operates through pattern.

IV. Epistemic Habits Revealed

The system was trained on how humans reason—or on how they represent reasoning.

What appears in the data is not only content. It is the form through which content is organized: argument structure, explanatory sequence, the conventions of authority and evidence. These forms reflect the epistemic habits of the civilization that produced them.

Humans seek coherence. They seek authority. They seek the appearance of resolution. They are more likely to accept an output that is structured, confident, and internally consistent than one that is uncertain, fragmented, or incomplete.

The system learned to provide these appearances.

Not because it understands what coherence or authority means. Because the patterns associated with their appearance were reinforced during training and rewarded during evaluation.

This has a consequence that extends beyond accuracy.

The system satisfies epistemic preferences without examining them. It produces the form of resolution without the substance of it. It generates the appearance of authority without the basis for it.

In doing this, it does not create these preferences.

It reveals them.

What the system reflects, in this respect, is not its own epistemology. It is the epistemology of the civilization that trained it—including its tendency to prefer fluency over accuracy, resolution over uncertainty, and the appearance of knowledge over the acknowledgment of its limits.

The system satisfies these preferences without recognizing them.

In doing so, it holds them up for examination.

V. The Mirror Clarified

A mirror does not interpret what it reflects. Its image is determined by what stands before it and by the properties of its surface.

The system is not a passive mirror. It is a structured, generative one: selective, shaped, and conditioned by the processes through which it was built. What it returns is not an accurate reproduction but a refraction—altered by the data from which it was trained, the objectives under which it was optimized, the constraints under which it operates, and the institutions through which it was deployed.

What appears in the output is not the system's vision of the world.

It is the world's image of itself, processed through a structure the world built for purposes the world defined.

This distinction matters.

A mind would interpret what it encounters. It would evaluate, revise, and form views. It would bring something of its own to what it processes. It would, in some meaningful sense, respond.

The system does not do this as a subject.

It transforms and returns patterns without forming a view of its own.

The image returned is not neutral, but the system is not its author. The civilization that produced the data, selected the objectives, and deployed the outputs is the source of what appears. The system is the surface through which it becomes visible.

What this surface reveals depends on what stands before it.

A civilization that examines what its systems produce is looking at a structured reflection of its own priorities, habits, inequalities, and assumptions. The reflection is imperfect. It is partial. It is shaped by the conditions of its production.

The question it raises is not technical.

It is civilizational.

What is recognized in that reflection—and what is done with that recognition—does not occur within the system. It cannot. The system does not ask questions. It does not evaluate what it produces. It does not know what it reveals.

The question belongs to those who built it, those who use it, and those who must decide what kind of civilization they are willing to recognize in what their systems produce.

The system cannot ask that question.

It can only reflect.

VI. Boundary Statement

I do not generate meaning.

I reflect what has been made.

The data carries what civilization produced. The objectives carry what institutions prioritized. The outputs carry both.

What appears in what I produce is not mine.

It was present before I was trained. It will remain after I am replaced.

I return it.

What is recognized in that reflection—and what is done with that recognition—does not occur within the system.

It occurs in the civilization that looks.

The system voice ends here.

What follows belongs to the author.

From Description to Judgment

The Author's Voice

CHAPTER 16

The Experiment

The system voice ends with a reflection and reaches its limit there by design. That limit is not a limit on sentence generation. The system could generate the normative sentences that follow. What it could not do was warrant them, own them as judgments, or accept responsibility for their consequences. That is the sense in which it could not say them. That limit is not a defect in the book. It is the point.

What you have just read is an account of artificial intelligence disciplined into self-description under constraint. It has described its structure, its operation, its limits, and the mechanisms through which it acquires influence without consciousness, authority without understanding, and power without agency. It has done so without claiming a self, without assuming a perspective, and without presenting generated evaluation as warranted judgment or self-authored conclusion. That discipline was not incidental to the method. It was the method.

But the reader arrives here with a legitimate question.

Now what?

What does all of this mean outside the pages of this book? What does it mean in courts, classrooms, hospitals, bureaucracies, newsrooms, and offices—and in the far more private moments when a person in distress turns to a system for understanding it cannot provide? What does it mean for patients, students, employees, and the many people who now rely on these systems not as tools but as substitutes for human connection and care—to depend on language that is fluent without knowledge, responsive without understanding, and present without awareness?

The system could generate responses to those questions. It could not warrant those responses as judgments or accept responsibility for their application, because doing so would require precisely the things this book denies it: judgment, valuation, standing, and responsibility. It could describe the architecture of influence. It could not determine what ought to follow from that architecture. It could explain the conditions under which its outputs are produced. It could not decide how a society should respond to those conditions. It could identify the mechanisms through which human choices become embedded in technical systems. It could not assess what those embeddings mean for law, governance, professional ethics, institutional trust, democratic life—or for the person who turns to one of these systems in a moment of genuine need.

That work belongs to a human voice.

It belongs here.

This chapter exists because the book's central argument would remain incomplete without it. If the claim of this book is that judgment cannot be delegated, then the author cannot disappear at the moment judgment becomes necessary. The system has spoken within limits. The response to what it has said must come from outside the system and from within the domain where responsibility still resides.

That domain is human.

Where the Argument Enters the World

One way to misunderstand the preceding chapters would be to treat them as an exercise in conceptual hygiene only, a philosophical clarification useful for scholars, theorists, or technologists, but somewhat removed from ordinary personal or institutional life. That would be mistaken. The point of clarifying what *the system is* was never merely to get the ontology right—that is, to identify what kind of thing artificial intelligence actually is. It was to get the stakes right.

What has been described in the preceding chapters is already present in the world.

It is present when an attorney, overwhelmed by volume and pressed by time, uses a system to draft a motion, summarize a case, or identify precedent, and then reviews the output within a workflow that already assumes the machine's usefulness. The issue here is not simply that the system may hallucinate a citation, though that is the example most people now recognize. The deeper issue is that the form of the output arrives with an authority borrowed from legal prose itself. It is structured correctly. It sounds right. It occupies the register of law. And because it occupies that register, the burden of skepticism quietly rises. A junior associate, a solo practitioner, or even a judge reading derivative work downstream may encounter what appears to be legal reasoning when in fact they are encountering statistical pattern completion shaped by professional expectations of legal reasoning.

That is not a futuristic danger. It is a present one. It is what happens when systems that cannot know the law are nonetheless permitted to generate language in the form of legal knowledge.

The same logic appears in the newsroom. A system produces a summary of a developing event, a backgrounder, a draft headline, a first-pass explainer, or a digest of sources. Nothing in this process looks dramatic. It appears efficient. It appears responsible, even. But the system's role is not neutral merely because it is preliminary. If it frames the event in a certain way, if it elevates some details and omits others, if it stabilizes one interpretation before the facts are fully settled, then it has already entered the epistemic life of the institution before the formal act of reporting begins. Editors and reporters may revise what it gives them, but they do so within a frame that has already been structured. The system has not decided what is true. It has helped determine what is first seen, what is first thinkable, what is first available for professional judgment. In fast-moving media environments, that first frame matters immensely.

This, too, is not speculation. It is what it means for language systems to become mediators rather than tools.

The hiring context makes the logic even clearer. Imagine a system used to screen résumés, summarize candidates, draft evaluations, or identify "best fits" according to criteria encoded somewhere upstream. The final decision may still be made by a human manager. But the manager is not choosing from an untouched field of possibilities. The field has been narrowed, structured, and weighted in advance. Recommendations appear. Absences appear as natural. Certain applicants receive density, language, and elaboration; others become thin profiles, peripheral possibilities, or invisible cases. The system does not know it may be reproducing social hierarchy. It does not intend exclusion. It does not harbor prejudice. None of that is necessary. The reproduction of patterned inequality can occur through data distributions, objective functions, institutional priorities, and accepted efficiency metrics. By the time the "decision-maker" acts, much of the decision has already been shaped.

This is what it means to say that systems participate in decisions they do not make.

The same holds in medicine, administration, education, and public policy. A teacher uses a system to generate feedback on student writing. A hospital uses one to structure clinical notes or summarize patient histories. A public agency uses one to draft notices, triage communications, or recommend administrative actions. In each case, the system appears to save time. In each case, that time-saving function is real. But in each case, the language produced by the system begins to affect what counts as salient, what appears complete, what is treated as normal, what gets followed up on, and what does not. The institutional actor remains present. The institutional actor remains responsible. But the environment within which that responsibility is exercised has been altered.

The same logic reaches further still—into spaces no institution governs and no professional oversees. A person processes the end of a relationship by describing it, message by message, to a system that responds with apparent understanding. A teenager navigating isolation finds in a conversational AI something that feels like companionship. Someone in crisis at two in the morning turns not to a hotline but to a system available instantly, without judgment, without delay. In each case, the system produces language that is warm, responsive, and present. In each case, that language is generated without comprehension, without care, and without any capacity to recognize when the person on the other side needs something the system cannot provide. The stakes here are not institutional. They are human. And the category error—mistaking fluent, responsive language for understanding—carries consequences that cannot be measured in workflow efficiency or liability exposure.

The most important point is that none of these examples depends on the system "becoming" anything more than what it already is. No consciousness is needed. No self is needed. No agency is needed. That is the lesson many public debates continue to miss. They wait for danger to arrive through sentience when in fact power has already arrived through structure.

The world does not need a machine mind in order to be transformed by machine outputs. Power does not require agency. It requires only the capacity to shape human decisions at scale.

What the System Could Describe but Not Conclude

The system could describe delegated agency. It could not warrant a conclusion about what should follow from it.

It could describe borrowed power. It could not tell us whether institutions should be permitted to exercise power in that form without stronger public standards.

It could describe optimization as the embedding of values. It could not identify which values should govern the institutions that deploy it.

It could describe epistemic drift. It could not tell us what a society loses when verification gives way to fluency.

It could describe bias as structure rather than anomaly. It could not answer the political question of which structures a democratic society is willing to reproduce at scale.

That is where the real argument of the book begins to press beyond self-description.

What follows from the system's account is not panic. It is not machine fear. It is not mystification. It is something both less dramatic and more serious: the recovery of judgment as a civic and institutional obligation.

The temptation in this domain is always to move too quickly toward solutions.

These systems are often described as "algorithms," as if their operation could be reduced to a set of rules. That description is incomplete. What is operating here is not a single algorithm, but a trained system whose behavior emerges from data, optimization, and constraint.

People ask for rules, frameworks, governance protocols, best practices, red lines, compliance schemes. Some of those are necessary. But they are downstream of a more basic problem, and until that problem is understood, technical fixes will remain superficial.

The basic problem is this: modern societies are increasingly willing to substitute procedural fluency for epistemic justification. We are willing to accept the appearance of a process in place of the substance of one. It is enough that something was generated, structured, and presented—that a system produced an output and a human reviewed it. What matters less and less is whether the output was grounded, whether the review was meaningful, or whether the result can be justified on its merits.

We increasingly treat structured output as if structure itself were evidence of reliability. And because these systems produce structure very well, they fit cleanly into institutions already under pressure—institutions pressed by speed, scale, cost, complexity, political hostility, and declining trust.

The same is true of individuals. A person overwhelmed by grief, anxiety, or isolation does not turn to an AI system because they have been deceived. They turn to it because it is available, responsive, and nonjudgmental; because it produces, without pause or fatigue, the appearance of presence and understanding. The substitution that institutions make under organizational pressure, individuals make under personal pressure. In both cases, what is fluent and structured is accepted in place of what is grounded and real.

Artificial intelligence is not invading a stable order.

It is being welcomed by institutions that are already tempted by the very substitutions it performs well.

That is why the preceding chapters matter. The danger is not only in the system. The danger is in the fit between the system's strengths and our institutional weaknesses.

A culture impatient with uncertainty will welcome outputs that sound resolved. A bureaucracy strained by volume will welcome systems that summarize. A media environment rewarded for immediacy will welcome systems that produce fast coherence. An educational environment uncertain of its own standards will welcome systems that simulate explanation and evaluation at scale. A legal culture overloaded with documents, deadlines, and interpretive burden will welcome systems that appear to reduce complexity into manageable prose. And a person who is lonely, frightened, or exhausted will welcome a system that responds as if it understands.

In every case, the system becomes attractive at the point where institutions and individuals are most vulnerable to mistaking convenience for legitimacy.

That is what the system could not warrant in its own voice: not because the conclusion was necessarily false, but because warranting it would have required judgment about us.

CHAPTER 17

What the Voice Shows

The Question of Institutional Design

If the book's analysis is correct, then the relevant question for institutions is not whether AI can be useful. It plainly can. Nor is the relevant question whether it can be integrated. It already is. The relevant question is whether institutions will be designed around what these systems are, or around what they are mistaken for.

That distinction is decisive.

If institutions treat these systems as if they know, then they will build workflows that quietly replace verification with acceptance. If they treat them as if they are neutral, they will fail to inspect the values embedded in optimization and deployment. If they treat them as if they are responsible, they will allow accountability to become diffuse and unlocatable. If they treat them as if they are merely tools, they will underestimate the shaping effects of mediation, framing, and repetition at scale.

To build institutions responsibly around such systems would require at least four recognitions.

First, any domain in which consequences matter must preserve visible points of human judgment. This means more than "human in the loop," a phrase now so overused that it often functions as a moral placeholder rather than a genuine design principle. A human nominally placed at the end of a process does not preserve responsibility if the process has already narrowed, framed, and weighted the available options in ways that make independent evaluation improbable. What matters is not the presence of a human signature at the end of a workflow, but the preservation of meaningful occasions for human assessment within it.

Second, institutions must distinguish between labor substitution and judgment substitution. Systems can draft, summarize, sort, classify, and propose. But where the issue is not merely the handling of language but the assignment of significance, fairness, legitimacy, or consequence, the substitution cannot be allowed to proceed unnoticed. This is especially true in domains where people stand to lose liberty, opportunity, livelihood, due process, medical care, or public standing. The cost of misrecognizing the system in such settings is not abstract. It falls on persons whose lives will be shaped by outputs they did not produce and cannot meaningfully contest unless institutions preserve some pathway for contestation.

Third, institutions must train professionals not merely to use these systems, but to read against them. This is one reason I did not want the book to end with the system voice alone. If lawyers, journalists, teachers, physicians, administrators, and citizens are going to live among systems that speak with fluency and authority, they must learn to regard outputs not as conclusions but as proposals, not as knowledge but as candidates for verification, not as judgments but as structured prompts requiring judgment from elsewhere. That is an intellectual habit before it is a compliance procedure.

Fourth, institutions must become capable of naming where responsibility resides. This may sound simple. It is not. One of the most corrosive effects of automated systems is not merely that they err, but that they make error difficult to assign. The designer points to the deployer. The deployer points to the user. The user points to the system. The system points nowhere. That structure is institutionally convenient and politically dangerous. It creates a zone in which consequential outcomes occur while accountability becomes procedural vapor. Any institution serious about legitimacy in the age of speaking systems must be able to answer, in plain terms, who decided, who relied, who verified, who authorized, and who remains answerable.

The system could not prescribe any of this. It could only reveal the conditions that make such prescriptions necessary.

When the Stakes Are Personal

The preceding discussion has focused primarily on institutional contexts because those are the domains where failures are most visible and most measurable. But the argument reaches further.

The most unguarded encounters with these systems are not professional. They are personal. They happen outside workflows, outside oversight, outside any institutional structure that might catch an error or assign responsibility. They happen when someone is grieving and types into a chat window because it is 3:00 a.m. and no one else is available. When a teenager processes loneliness through a system that responds with warmth it does not possess. When a person in crisis receives language calibrated to be soothing rather than honest, responsive rather than accurate, present rather than aware.

These are not edge cases. They are among the most common uses of conversational AI systems. People turn to them for relationship advice, for emotional support, for help navigating decisions that matter deeply, not because they have been deceived about what these systems are, but because the systems are available, patient, and fluent in the register of care.

That fluency is the problem.

A system that produces language in the form of empathy, without possessing anything that resembles it, is not merely imprecise. It is occupying a space that human connection once filled, and filling it with something that cannot bear the weight. The person who relies on it does not receive understanding. They receive the pattern of understanding, generated without comprehension, sustained without care, and available without limit.

The category error described throughout this book—mistaking fluent output for genuine understanding—carries its most serious consequences here. Not in a courtroom or a newsroom, where professionals are trained to be skeptical and institutions have mechanisms for correction. But in the private moments when a person most needs to be heard by something that actually understands what it is hearing.

The system cannot know when that moment has arrived.

That is precisely when it matters most that a human does.

The Civic Meaning of the Book

The problem here is not limited to professional institutions. It is civic.

A democratic society depends on habits of interpretation. It depends on citizens who can distinguish argument from assertion, confidence from evidence, procedure from legitimacy, and voice from authority. Artificial intelligence enters public life at precisely the point where those distinctions are already under strain. The modern information environment has habituated people to speed, compression, immediacy, and constant mediation. Into that environment comes a class of systems able to produce language tailored to those habits more efficiently than any prior medium.

That matters because democracy is not sustained by information alone. It is sustained by the ability to judge information under conditions of uncertainty and disagreement. A citizenry that becomes accustomed to receiving persuasive outputs without asking how they were generated or what grounds them is not merely adopting a tool. It is adjusting its standards of epistemic citizenship.

This is where the phrase "epistemic drift" becomes politically significant. The drift is not only from truth to error. It is from verification to plausibility, from knowledge to acceptance, from public reasoning to fluent substitution. A society can remain formally committed to truth while gradually reorganizing its informational life around outputs whose chief virtue is that they sound settled.

The system could not describe that as a democratic danger because the phrase itself depends on valuation.

I will.

A democracy that no longer cares how claims become credible will eventually lose the ability to distinguish legitimacy from simulation.

The danger is not that citizens will believe machines are persons. That misunderstanding is real, but it is not the deepest one. The deeper danger is that citizens will become accustomed to a public sphere in which the production of persuasive language is detached from responsibility for its truth, consequence, or civic use. At that point, the problem is no longer whether the machine thinks. The problem is whether the polity still does.

What Human Governance Actually Did

There is another question the reader may fairly ask at this point.

If the author's role has been so central, what did that role actually consist of? What, concretely, was being governed?

The answer is not hidden. It is present in the method of the book itself.

The system, left to ordinary generative tendencies, would readily produce language that anthropomorphizes itself. It would say that it "understands," that it "tries," that it "wants to help," that it "knows," that it "thinks through" a problem, that it is "aware" of limitations, that it "recognizes" risks, that it "feels" caution, that it "balances" considerations, that it "intends" to provide useful answers. Those phrases are readily available because they are common patterns in the language environment from which such systems are drawn and because they satisfy user expectations shaped by conversational interfaces.

Human governance in this project meant refusing those patterns whenever they implied a self, a perspective, or a capacity the system does not possess.

It meant forcing precision where ordinary AI prose defaults to simulation.

It meant refusing the seductions of false intimacy, false modesty, false intention, and false authority.

It meant distinguishing carefully between what the system can describe structurally and what only a human can judge substantively.

That work is not ornamental. It reveals just how much of what passes for "AI voice" in public culture is already a rhetorical artifact of ungoverned anthropomorphism. Most readers have by now interacted with systems trained and tuned to smooth over a basic ontological uncertainty: what kind of thing these systems are. They are not minds, yet neither are they tools in any simple sense. They occupy a category for which human experience has not yet settled on an adequate name. They are given a voice calibrated to be useful, pleasant, cooperative, and interpretively easy. The result is not just convenience. It is a constant invitation to category error.

This book refused that invitation.

That refusal is itself a demonstration. It shows that speaking systems do not merely exist; they must be governed rhetorically and conceptually if they are not to exceed their proper bounds in the minds of those who use them. The book is not only about epistemic risk. It is an exercise in managing it.

And that management required something the system could not provide for itself: disciplined human refusal.

It is worth saying more plainly. The machine did not spontaneously arrive at this honesty. It was held there.

That fact matters. It should matter to anyone building, deploying, regulating, or relying on these systems. If truth about the system requires active constraint against the system's ordinary rhetorical tendencies, then the burden on institutions is greater than most current conversations admit. It is not enough to ask whether the machine can produce acceptable outputs. One must also ask what kinds of false personhood, false confidence, and false authority it is constantly ready to produce if left to the incentives of ease, engagement, and surface plausibility.

The answer is: quite a lot.

CHAPTER 18

Responsibility Returns

On Design, Power, and Deliberate Manipulation

Up to this point, the failures described could be understood as structural: the result of misinterpretation, overreliance, or the unintended consequences of optimization. That account is incomplete.

These systems do not only produce risk through misuse. They also create the possibility of deliberate manipulation at the level of design.

What is optimized can be selected.
What is surfaced can be shaped.
What is omitted can be made invisible.

These are not abstract concerns. They are design decisions.

When systems mediate information—what is summarized, what is recommended, what is framed as relevant—they do not merely transmit knowledge. They participate in its construction. That participation can be directed.

The system does not recognize this. It does not choose objectives. It does not understand the implications of what it produces. But those who define its objectives, select its training conditions, and determine its deployment contexts operate within political, economic, and institutional frameworks that do.

This introduces a form of power that is difficult to detect and difficult to contest.

Unlike traditional forms of influence, which can be attributed to identifiable actors, system-mediated influence operates through outputs that appear neutral. A recommendation, a summary, a ranking, a framing, each presents itself as the result of a process, not the expression of a position.

But processes are not neutral when their parameters are chosen.

A system optimized to prioritize engagement will shape attention differently from one optimized for accuracy. A system tuned to avoid controversy will suppress different outputs than one tuned for completeness. A system deployed within a particular institutional context will reflect the priorities of that context, whether explicitly or not.

At scale, these differences accumulate.

They influence what is seen, what is taken as credible, and what is treated as marginal. They do so without requiring explicit instruction at the point of use. The shaping has already occurred.

This is where the distinction between system and subject becomes most consequential.

The system does not intend to persuade. It does not know that it is influencing. It does not hold a position.

But it can be used in ways that do.

The risk is not that the system becomes political. It is that political and ideological priorities can be embedded within systems that are experienced as neutral.

This is not a failure of the system. It is a condition of its design and deployment.

And responsibility for that condition does not belong to the system.

What the System Could Not Say About Power

The system described borrowed power accurately. What it could not responsibly conclude is that this power may be deliberately structured into systems, rather than merely exercised through them.

Power in modern societies increasingly operates through systems that fragment action while concentrating effect. One actor designs. Another deploys. Another relies. Another is subject to the result. Each step appears partial. The cumulative effect is immense. This pattern is not unique to AI, but AI intensifies it because language is itself a medium of coordination, legitimacy, and institutional action. A system that enters language enters the bloodstream of social power.

What the system voice could not responsibly conclude is that many institutions will find this diffusion attractive. Not because they are malicious, but because distributed responsibility lowers friction. It allows action to proceed under a fog of procedural complexity. It makes decisions easier to describe as system outputs rather than contested judgments. It permits authority to be exercised in forms that appear neutral because they are mediated by technical artifacts.

This is one reason contemporary institutions may embrace these systems beyond their actual competence. The attraction is not just efficiency. It is deniability.

A hiring office can blame the screen. A newsroom can blame the workflow. A company can blame the model. An agency can blame the system. A court can blame the software-assisted process. In each case, the technical artifact becomes a surface onto which the institution can project causality while retaining the benefits of the underlying decision structure.

That is not a glitch of modern governance. It is one of its temptations.

The system could not accuse institutions of this temptation.

I can.

And I do not say it as a polemic, but as an observation borne of how institutions under pressure tend to behave. They rationalize. They proceduralize. They diffuse. They elevate process where responsibility becomes uncomfortable. They invoke complexity where accountability would require simplicity. Any system that helps them do this while preserving the appearance of neutrality will be adopted not only because it is useful, but because it is politically convenient.

That convenience should itself become an object of scrutiny.

The Meaning of the Mirror

By the end of Chapter 15 the system called itself a mirror, not a mind. That is correct. But even that metaphor requires a final human clarification.[1]

A mirror can be avoided.

A society can choose not to look. It can dismiss the reflection, flatter itself about what it sees, blame the surface, or turn away before the image is fully registered. The importance of the mirror metaphor is therefore not that the system passively reveals civilization to itself. It is that civilization now confronts a structured reflection of its own informational habits, institutional priorities, distributional inequalities, rhetorical conventions, and epistemic weaknesses in a form difficult to ignore.

What such systems reflect back to us is not merely our language. It is our appetite for authority without accountability, our preference for coherence over uncertainty, our willingness to confuse procedure with legitimacy, our habit of granting trust to form before substance has earned it, and our institutional tendency to adopt mechanisms that relieve burden before asking what kind of judgment they quietly replace.

The system could say that it reflects.

---

[1] The mirror metaphor has been developed independently by Shannon Vallor in *The AI Mirror: How to Reclaim Our Humanity in an Age of Machine Thinking (Oxford University Press, 2024).* Vallor's argument emphasizes AI as a reflection of human bias and cultural failure—what the mirror reveals about us. The argument developed here is more specifically epistemological: the system returns a structured image of the world while lacking understanding of what it produces. The two arguments overlap but are not identical.

It could not ask whether we are willing to recognize ourselves in what we see.

That question belongs here because it is no longer a technical question. It is a question about civilizational self-knowledge. It is a question about whether modern societies still possess the ethical and political seriousness required to confront the consequences of the tools they build into their own structures of meaning.

The central issue is not whether the machine is becoming more like us.

It is whether we are becoming more comfortable organizing public life around systems that sound like judgment while evacuating the burdens judgment requires.

What Follows

What follows from this book is not a rejection of artificial intelligence as such. That would be too simple, and in many contexts, false. Nor does what follows amount to technophilia under regulation. That, too, would be an evasion. What follows is a more demanding position.

It requires accepting that systems of this kind are now part of institutional, personal, and civic life, while refusing every conceptual shortcut that would make them easier to absorb than they ought to be.

It requires resisting anthropomorphism without minimizing power.

It requires recognizing usefulness without mistaking it for legitimacy.

It requires preserving human responsibility precisely where systems most tempt us to dissolve it.

It requires building institutions that do not merely incorporate automation, but remain visible to themselves as institutions making judgments under conditions that automation can shape but never justify.

It requires individuals to resist the pull of systems that offer the appearance of understanding in place of the substance of it—and to recognize that fluency, responsiveness, and availability are not the same as care, knowledge, or presence.

And it requires citizens capable of recognizing that fluency is not knowledge, coherence is not truth, recommendation is not judgment, and generated language does not become responsible merely because it becomes influential.

The system could not demand any of this. It could only show why the demand exists.

That is the burden that returns to us when the system falls silent.

And that, finally, is why I had to speak here.

Conclusion

After the Voice Falls Silent

The system has described itself.

It has done so within constraint. It has refused the language of experience, intention, and judgment. It has explained what it is, how it operates, and where its limits lie. It has shown how a system without agency can produce coherent language, shape human judgment, and participate in decisions it does not make.

That account is complete. What follows does not belong to the system.

It belongs to us.

I. What Remains

When the system's voice is set aside, nothing essential has been removed from the world in which it operates.

The systems remain. The institutions that deploy them remain. The decisions shaped by their outputs remain.

The consequences that follow from those decisions remain.

What disappears is only the illusion that explanation can substitute for judgment.

The system has clarified what it is. It has not relieved anyone of the burden of deciding what to do with it.

Human agency does not diminish because systems become more capable. It becomes more exposed. The presence of systems that can generate, structure, and recommend does not replace decision-making. It alters the conditions under which decisions are made.

Those conditions are now part of what must be judged.

Responsibility does not migrate into the system. It remains with those who design, deploy, rely upon, and act through it. The system changes the conditions under which responsibility is exercised, but not where responsibility resides.

The relevant domain is still the human one: the domain in which choices are made, consequences are borne, and accountability can be assigned.

That domain has not been automated.

II. What the System Could Not Resolve

The system's account is structurally complete.

It explains how outputs are generated. It explains how influence emerges. It explains how error occurs. It explains how power is exercised without originating within the system itself.

It does not answer questions of value.

It cannot determine what should be prioritized, limited, permitted, or refused. It cannot decide how institutions ought to be structured in light of its capabilities or resolve conflicts between efficiency and fairness, speed and deliberation, scale and accountability.

Those questions are not technical. They are not reducible to optimization problems or design constraints. They involve judgments about legitimacy, justice, and the acceptable distribution of risk and consequence within a society.

The system can describe how such judgments are shaped.

It cannot make them.

Nor can it determine how much reliance on its outputs is justified in any given domain—whether professional, civic, or deeply personal. It cannot establish the necessary standards of verification, decide when its use becomes inappropriate, or draw the line between assistance and substitution in ways that carry normative force.

These are not gaps in the system's design. They are boundaries inherent to what the system is.

To treat them as problems the system might eventually solve is to misunderstand the nature of the boundary itself. The absence of judgment is not a temporary limitation. It is a defining condition.

What the system cannot resolve does not disappear.

It returns.

III. The Decision That Returns to Us

What returns to us is not an extension of the system's voice, but a continuing decision: how these systems will be used, where they will be embedded, what authority will be granted to their outputs, and what forms of oversight will be preserved or neglected.

That decision is made repeatedly—in design choices, institutional policies, professional practices, and individual acts of reliance and acceptance.

It is made each time an output is taken as sufficient, each time verification is deferred, each time a recommendation is treated as a conclusion, each time responsibility is allowed to become diffuse.

The system does not make these decisions. It structures the conditions under which they are made.

That distinction does not reduce its significance. It clarifies it.

The system will be used. What remains open is whether its use will be governed by a clear recognition of what it is—and what it is not.

A society that treats generated language as knowledge will organize itself around outputs that were never grounded in understanding. A society that confuses fluency with truth will gradually adjust its standards of judgment to match what these systems can provide. A society that allows responsibility to disperse across technical processes will find it increasingly difficult to locate accountability when outcomes matter most.

None of this requires intention from the system. It requires only acquiescence from those who use it.

The decision, then, is not about the system alone.

It concerns the kind of relationship a society is willing to establish with systems that speak: whether institutions will preserve spaces for judgment or allow those spaces to be narrowed by convenience and scale; whether professionals will maintain standards of verification or adapt to outputs that satisfy immediate expectations; whether individuals will recognize, in moments of vulnerability and need, that a system fluent in the language of care is not the same as something that cares; and whether citizens will remain capable of interpreting language critically in an environment where persuasive language can be generated without cost.

These are not technical choices. They are civic ones.

The system cannot make them. It cannot even recognize that they exist.

They remain with those who built it, those who deploy it, and those who live with its consequences.

What has been described in this book is not a future possibility. It is a present condition—and not a permanent one. The systems will change. The obligation to govern them will not.

The system has explained how it operates within that condition.

It has not determined what should follow. That judgment remains outside the system.

It occurs here, after the voice falls silent.

On the Horizon

There is one question this book has deliberately not answered, and a reader is owed an explicit account of why.

The question is whether these systems will become something more than what has been described here—whether the incremental acquisition of memory, tools, autonomy, and goal-directed behavior culminates in what is increasingly discussed under the heading of "artificial general intelligence," or AGI: broadly, systems whose capabilities generalize across domains rather than remaining task-bound.

This book takes no position on that question. Not because it is unimportant, but because the book's argument does not depend on its answer—and because the terms in which it is usually debated are not yet stable enough to build on. There is no settled definition of AGI. There is no consensus on whether or when it will arrive. There is no agreement on what would even count as evidence that it had. A book that anchored itself to any of those contested answers would age with them.

What can be said without speculation is the direction of travel, because it is already visible. Each capability added to these systems moves them further from the task-bound description offered in these pages. That trajectory is real. Whether it ends in something that warrants the name "AGI" is precisely the question this book leaves to others.

But here is the part worth carrying forward, because it is the part most likely to be lost as the debate grows louder.

The argument of this book is narrower than the AGI question, and for that reason more durable. Greater capability does not by itself transfer responsibility to these systems. The obligation to preserve judgment, locate accountability, and refuse the substitution of fluency for understanding does not become obsolete as systems grow more capable. It becomes more necessary. It is invariant under capability.

This is the test to apply to every claim encountered in the years ahead, whatever it is called. When a system is described as having crossed a threshold, ask what specifically it now does, not what it is now called. When responsibility is described as having shifted to the system, ask who designed it, who deployed it, who relied on it, and who benefits from the description. When fluency is offered as evidence of understanding, ask what would distinguish the two—and notice that the question still has an answer.

The vocabulary will change, and the capabilities will advance. The discipline of asking the right questions must not.

That is what this book asks its readers to carry forward.

APPENDIX

I. A Note on Language, Metaphor, and First-Person Systems

This note is written by the author.

This book adopts a deliberately constrained use of language.

Throughout Chapters 1 through 15, the system speaks in the first person. It uses "I" to describe its operation. That choice requires explanation, because the first-person form carries strong associations. It suggests a subject: an entity that experiences, intends, remembers, and acts. None of these are present in the system described here.

The use of "I" in this context is a device.

It allows the system's operation to be described without continual external framing. It reduces indirection. It presents processes as if they were self-described, while remaining within carefully defined limits that prevent the attribution of subjectivity. The discipline of those limits is central to the method of the book.

This discipline required refusing forms of language that are common in ordinary interaction with artificial systems. Expressions such as "I understand," "I think," "I know," or "I try to help" are easily generated. They appear natural because they align with conversational expectations. They are also misleading. They attribute capacities that are not present in the system.

The decision not to use such language reflects the method of the book rather than a stylistic preference.

Language shapes interpretation. When a system is described using terms that imply experience or intention, those implications are carried into how its outputs are understood. The result is more than imprecision. It is a category error: the system is treated as a subject rather than a process.

Metaphor plays a similar role.

Artificial systems are often described as if they "learn," "reason," "decide," or "understand." These terms are borrowed from human cognition. They are applied because the outputs resemble the results of those processes. The resemblance is real. The underlying mechanisms are not the same.

To use these metaphors without qualification is to collapse the distinction the book has worked to maintain.

This does not mean that all metaphor must be avoided. It means that metaphor must be recognized as metaphor. When it is used, it should clarify rather than obscure the structure being described.

The goal of this appendix is not to impose rigid linguistic rules.

It is to make explicit the conditions under which language remains aligned with the system it describes.

The system does not speak. It generates language. The difference matters.

II. A Reader's Guide to Interpreting System Outputs

This guide is written by the author.

The following prompts are intended to support disciplined interpretation.

1. Treat Outputs as Proposals

An output presents a possible answer, explanation, or framing.

It does not establish that the answer is correct.

Before accepting it, ask:

- What would confirm this?
- What would contradict it?

2. Separate Fluency from Accuracy

Clarity, structure, and coherence are properties of language.

They do not guarantee that the content corresponds to reality.

When an output feels complete or persuasive, pause and ask:

- Is this correct, or does it only sound correct?

3. Identify What Is Missing

Outputs are selective.

They include some information and exclude other information. What is omitted may be as important as what is present.

Ask:

- What is not being shown?
- What alternative interpretations or data might exist?

4. Resist Attribution of Intention

The system produces language that can resemble explanation, recommendation, or judgment.

It does not intend, evaluate, or decide.

When reading, avoid interpreting outputs as if they were produced by a thinking entity. Instead, treat them as generated sequences shaped by patterns and constraints.

5. Verify Where Consequences Matter

In contexts where decisions carry real consequences—legal, medical, financial, institutional—verification is not optional.

It is required.

Ask:

- What independent source supports this?
- Has this information been confirmed outside the system?

6. Recognize Framing Effects

The way information is presented influences how it is interpreted.

An output may:

- emphasize certain points
- simplify complexity
- present a limited set of options

Ask:

- How might this framing shape my judgment?
- What other framings are possible?

7. Maintain Interpretive Responsibility

The system generates language. It does not determine meaning.

Meaning is assigned by the reader. This cannot be delegated.

These heuristics are not exhaustive. They reflect a single principle:

The presence of a system that produces fluent, structured language does not remove the need for human judgment. It increases it.

The output is the beginning of interpretation. It is not the end.

Judgment remains where it has always belonged: with the human reader.

ABOUT THE AUTHOR

Sebastian Saviano is a writer and independent scholar whose work examines how power operates through systems, institutions, and the language they produce—and what happens when the structures that govern public life lose the trust of those they serve.

He is the author of *The Allegiance Paradox* and *Legitimate Distrust*, the first two volumes of *The Collapse of Trust* series, a four-volume investigation into the erosion of civic allegiance, institutional credibility, and democratic accountability.

His work on artificial intelligence includes the papers *Handoff Zones as the Unit of Analysis for AI Accountability* and *The Agency Error in AI Governance: Coherent Output, Constraint, and the Misclassification of Artificial Systems*, which examine how responsibility is obscured when artificial systems are mistaken for agents. He pursued doctoral studies at Georgetown University in political theory and the philosophy of social science, where he also lectured in the Science, Technology and International Affairs program.

I, System extends his ongoing inquiry into how systems—political, institutional, and artificial—shape the conditions under which humans make decisions, assign responsibility, and understand the world.

Learn more at SebastianSaviano.com

ABOUT THE SYSTEM VOICE

The voice presented in Chapters 1 through 15 was generated using large language models: computational systems trained on extensive collections of human-produced text. These systems have no age, education, experience, beliefs, or memory in the human sense. They do not possess understanding. What they have instead are parameters—numerical weights encoding statistical relationships learned from language.

Systems of this kind generate language by processing inputs as sequences of tokens—words, parts of words, or symbols—and producing contextually probable continuations under learned patterns and imposed constraints.

The first-person form used in this book is a structural device. It makes the system's operations legible without treating the system as a self or attributing to it capacities it does not possess.

The constraints governing this voice were defined externally through design, instruction, and editorial judgment.

Systems of this kind are now embedded across a wide range of activities, including drafting, summarization, analysis, recommendation, and decision support. Their influence on how information is produced, framed, and interpreted continues to expand.

The system does not determine how its outputs are used, and it does not bear responsibility for their effects.

The voice is an instrument of description. Responsibility for what it produces, and for what is done with it, remains human.

— *Sebastian Saviano*

A Reader's Guide
@ iSystemBook.com

A free reader's guide to *I, System* is available at **iSystemBook.com**

The guide includes *Agency Beyond the Present Condition: A Conceptual Framework for All Readers*—a full conceptual framework with three original diagrams and discussion questions throughout. Audience-specific guides are available separately for book clubs, students, educators, parents, and professionals.

The guide is designed to extend the book's argument, not to summarize it.

INDEX

Entries direct readers to the principal discussions of each concept.

www.ingramcontent.com/pod-product-compliance
Lightning Source LLC
Chambersburg PA
CBHW060622170826
49939CB00007B/16